LOGIT MODELS FROM ECONOMICS AND OTHER FIELDS

J. S. CRAMER
University of Amsterdam and Tinbergen Institute

CAMBRIDGE UNIVERSITY PRESS
Cambridge, New York, Melbourne, Madrid, Cape Town,
Singapore, São Paulo, Delhi, Tokyo, Mexico City

Cambridge University Press
The Edinburgh Building, Cambridge CB2 8RU, UK

Published in the United States of America by Cambridge University Press, New York

www.cambridge.org
Information on this title: www.cambridge.org/9780521188036

First published 2003
First paperback edition 2011

A catalogue record for this publication is available from the British Library

Library of Congress Cataloging in Publication Data

Cramer, J. S. (Jan Salomon), 1928–
Logit models from economics and other fields/J. S. Cramer.
p. cm.
Includes bibliographical references and index.
ISBN 0 521 81588 6
1. Econometric models. 2. Logits. I. Title.
HB141.C722 2003
330.015195 – dc21   2002041450

ISBN 978-0-521-81588-8 Hardback
ISBN 978-0-521-18803-6 Paperback

Contents

List of figures

List of tables

Preface

For well over a decade statistical program packages have been providing standard routines for fitting probit and logit models, and these methods have become commonplace tools of applied statistical research. The logit model is the more versatile of the two: its simple and elegant analytical properties permit its use in widely different contexts and for a variety of purposes. This monograph treats logistic regression as the core of a number of such variations and generalizations. Its purpose is to present several widely used models that are based on the logit transformation and to answer the questions that arise in the practice of empirical research. It explains the theoretical background of these models, their estimation and some further statistical analysis, the interpretation of the results, and their practical applications, all at an elementary level. I assume that readers are familiar with ordinary linear regression and with the estimation theory and matrix algebra that go with it.

Parts of the book are taken from *The Logit Model: an Introduction for Economists*, published in 1989 by Edward Arnold. One of the things I have learned since then is that several varieties of logit analysis have been developed independently, in almost perfect isolation, in various branches of biology, medicine, economics, and still other disciplines. In each field practitioners use distinct approaches, interpretations and terminologies of their own. I have tried to overcome these differences and to do justice to the main developments in other fields, but this will not conceal that my own background and inspiration are in econometrics.

I have also recorded the course of the discovery of the logit by former generations and some highlights of its subsequent development. This material of historical and nostalgic interest is collected in the last chapter, which readers can easily avoid.

I have benefited from the advice of Heinz Neudecker, Hans van Ophem and Jan Sandee, even when I did not follow it, and from the technical support of Jeroen Roodhart and Ruud Koning.

J. S. Cramer

1
Introduction

1.1 The role of the logit model

Logit analysis is in many ways the natural complement of ordinary linear regression whenever the regressand is not a continuous variable but a state which may or may not hold, or a category in a given classification. When such discrete variables occur among the independent variables or regressors of a regression equation, they are dealt with by the introduction of one or several $(0,1)$ dummy variables; but when the *dependent* variable belongs to this type, the regression model breaks down. Logit analysis or logistic regression (which are two names for the same method) provides a ready alternative. At first sight it is quite different from the familiar linear regression model, and slightly frightening by its apparent complexity; yet the two models have much in common.

First, both models belong to the realm of causal relations, as opposed to statistical association; there is a clear *a priori* asymmetry between the oddly named independent variables, the regressors or covariates, which are the explanatory variables or determinants, and the dependent variable or outcome. Both models were initially designed for the analysis of experimental data, or at least for data where the direction of causation is not in doubt. In interpreting empirical applications it is often helpful to bear these origins in mind.

Within this causal context, the ordinary linear regression model offers a crude but almost universal framework for empirical analysis. Admittedly it is often no more than a simplified approximation to something else that would presumably be better; but it does serve, within its limitations, for empirical screening of the evidence. Logistic regression can be used in quite the same way for categorical phenomena.

There are of course also differences. Unlike regression, the logit model

1

permits of a specific economic interpretation in terms of utility maximization in situations of discrete choice. Among economists this confers
a higher status on the model than that of a convenient empirical device.
And there is a subtle distinction in that the ordinary regression model
requires a disturbance term which is stuck on to the systematic part as a
necessary nuisance, while in the logit model the random character of the
outcome is an integral part of the initial specification. Together with the
probit model, the logit model belongs to the class of *probability models*
that determine discrete probabilities over a limited number of possible
outcomes.

Finally, like the regression model, the logit model permits of all sorts of
extensions and of quite sophisticated variants. Some of these are touched
upon in the later chapters, but the present text is mainly concerned
with plain logistic regression as a convenient vehicle for studying the
determination of categorical variables.

A survey of the literature will show that a number of different varieties of the same model have been developed in almost perfect isolation
in various disciplines such as biology (toxicology), medicine (epidemiology) and economics (econometrics). This has given rise to separate and
distinct paradigms which share a common statistical core, but which employ different approaches, terminologies and interpretations, since they
deal with different types of data and pursue different ends. Even when
the application of the technique has become a mechanical routine, the
original justificatory arguments still linger at the back of the practitioners' minds and condition their views. In the present text we follow the
approach which originated in the bio-assay of toxicology and was later
adopted (and developed further) in certain branches of economics and
econometrics. We shall however try to provide links to the work in other
fields wherever this is appropriate, and the reader is encouraged to follow
these up.

1.2 Plan of the book and further reading

The book consists of nine chapters. Chapter 2 presents the central model
that sets the course. Consideration of a single attribute gives rise to the
binary model, and this simple vehicle carries the overwhelming majority
of practical applications. Chapter 3 deals at some length with its estimation by what is now the standard method of maximum likelihood;
while most readers will rely on program packages for their calculations,
it is important that they understand the method since it determines the

properties of the resulting estimates. Chapter 3 also provides a practical illustration. Chapter 4 deals with some statistical tests and the assessment of fit, and Chapter 5 with defects like outliers, misclassified outcomes and omitted variables. At this stage just over half of the book has dealt exclusively with the simple case of a binary analysis in a random sample. In Chapter 6 we consider the analysis of separate samples, but still only of a pair. It is only in the next two chapters that the treatment is widened to more than two possible outcomes. Chapter 7 is devoted to the standard multinomial model, which is a fairly straightforward generalization of the binary model; the particular variety of multinomial models known as random utility models, which is an economic specialty, is the subject of Chapter 8. Some but not all of the embellishments of the binary model in Chapters 4, 5 and 6 carry easily over to the multinomial case. Finally Chapter 9 gives a brief account of the history of the subject, with special reference to the approach adopted here.

Since this is after all a slim book, designed for newcomers to the subject, readers are expected to skim through the entire text, and then to return when the need arises to the bits they can use, or – even better – to continue at once with further reading. We have already noted the diversity of parallel but quite separate developments of essentially the same subject in a number of disciplines. Apart from natural differences in the type of data under consideration and in the ends that are pursued, further differences of style have arisen in the course of this development. No discipline is satisfied with establishing empirical regularities. In epidemiology and medicine the tendency is towards simplicity of technique, but statistical associations, even if they are as strong as between cigarette smoking and lung cancer, must be complemented by a reconstruction of the physiological process before they are fully accepted. Economists and econometricians on the other hand never leave well alone, favour all sorts of complications of the statistical model, and believe that the validity of empirical results is enhanced if they can be understood in terms of optimizing behaviour. Each discipline thus has a paradigm of its own that supports the same statistical technique, with huge differences in approach, terminology and interpretation. These differences are at times exaggerated to almost ideological dimensions and they have become an obstacle to open communication between scholars from different disciplines. The reader's outlook will be considerably widened by looking with an open mind at one or two texts from an alien discipline. For this purpose we recommend such varied sources as the classic book on bio-assay of Finney (1971) (first published in 1947), or the up-to-date monograph on logistic regression from the epidemiological perspective by Hosmer and Lemeshow (2000). For the purely statistical view the reader can turn to Cox and Snell (1989), to the much more general text of Mc-Cullagh and Nelder (1989), or to the treatise on categorical variables by Agresti (1996). The survey article on case–control studies by Breslow (1996) reflects the practice of medical and epidemiological research. For the econometric approach one should read the early survey article of Amemiya (1981) or, for a rigorous treatment, Chapter 9 of his textbook of 1985. Another text from econometrics is Maddala's wide ranging survey of a whole menagerie of related models (1983), or, with a much more theoretical slant, the book of Gourieroux (2000). An introductory text that conveys the flavour of the use of the logit model in the social sciences is Menard (1995). In the book by Franses and Paap (2001) these techniques are presented as part of a much broader range with a view to marketing applications. For early economic applications we refer to the handbook of discrete choice in transportation studies of Ben-Akiva and Lerman (1987); a recent record of the achievements in this tradition is McFadden's address of acceptance of the Nobel prize (2001). Pudney (1989) gives a rigorous survey of advanced micro-economics with equal attention to the theory and to empirical issues.

This list is by no means complete, and there are also further specializations within each field: the current literature in learned journals shows a separate development of econometrics in marketing and in finance. Readers should browse for themselves to keep abreast of these advances.

1.3 Program packages and a data set

Maximum likelihood estimation is by now the accepted standard method of estimation, and for the simple binary model and the standard multinomial model this is included as a simple routine in many program packages. One of the first to do so was the BMDP package (for BIOMEDICAL DATA PROCESSING), in the late 1970s, but by now logit and probit routines are a common part of general statistical packages like SAS, SPSS and STATA. They are also found in programs of econometric inspiration, like TSP (Time Series Processing), LIMDEP (specifically aimed at the wider class of Limited Dependent Variables) and E-VIEWS (so far binary models only). All these routines will provide coefficient estimates with their standard errors and a varying assortment of diagnostic statistics. Most illustrations in this book have been produced by LOGITJD, an early fore-runner of the logit module that is now part of the econometric program package PCGIVE. It is useless to give further technical details of this and other packages as they are continually being revised and updated.

Many programs, however, do not permit the immediate calculation of specific magnitudes like the Hosmer–Lemeshow goodness-of-fit test statistic, nor do they all readily permit the estimation of even quite mild variations of the standard model, like the logit model with allowance for misclassification or the nested logit model. While the standard routines can sometimes with great ingenuity be adapted to produce nonstandard results, analysts who wish to explore new avenues are much better off writing specific programs of their own in programming languages like GAUSS or OX. Suitable short-cut procedures for maximum likelihood estimation are available in either language and these can be embedded in programs that suit the particular wishes of the analyst. By now these programming languages are quite user-friendly, and the effort of learning to use them is amply rewarded by the freedom to trim the statistical analysis to one's personal tastes.

Many program packages and some textbooks come with sample data sets that readers can use for exercises. In this book we make repeated use of a data set on private car ownership of Dutch households in 1980, and since 2000 this has been made available to users of the PCGIVE program package. It is now available to all readers of this text, who can obtain it from the Cambridge University Press website. The address is http://publishing.cambridge.org/resources/0521815886/.

The data come from a household budget survey among Dutch household held in 1980 by the then Dutch Central Bureau of Statistics (now Statistics Netherlands). This survey recorded extensive and detailed information about income and expenditure of over 2800 households. The survey is unusual in that it contains a great deal of information about the cars at the disposal of the households, with a distinction between private cars and business cars that are primarily used for business and professional purposes. Although the data are by now out of date as a source of information about car ownership, they are well suited to demonstrate various models (as in this book) and for trying out new statistical techniques (as in a number of other studies). As a rule, Statistics Netherlands, like all statistical agencies, is very reluctant to release individual survey records to third parties, in view of the disclosure risk, that is the risk that individual respondents can be identified. In the present case a welcome exception was made since the information, which is anyhow severely limited, is over 20 years old.

We use only a small fraction of the rich material of the budget survey, namely the information about car ownership, income, family size, urbanization and age. The data set consists of 2820 records, one for each household, with six variables. In the order of the dataset (which differs from the order of the analyses in this book) these are:

- PRIVATE CAR OWNERSHIP status in four categories, numbered from 0 to 3, namely *none, used, new* (for one used or new car respectively) and *more*. Private cars are all cars at the disposal of the households that are not business cars (see below).

- INC, income per equivalent adult in Dutch guilders per annum.

- SIZE, household size, measured by the number of equivalent adults. This is calculated by counting the first adult as 1, other adults as 0.7, and children as 0.5.

- AGE, the age of the head of household, measured by five-year classes, starting with the class 'below 20'.

- URBA, the degree of urbanization, measured on a six-point scale from countryside (1) to city (6).

- BUSCAR, a $(0,1)$ dummy variable for the presence of a business car in the household. A business car is a car that is primarily used for business or professional purposes, regardless of whether it is paid for wholly or in part by the employer or whether its costs are tax-deductible.

In all analyses in this book we follow the common usage of taking the logarithm of income and, since it is closely related to this, of size as well, denoting the transformed variables by LINC and LSIZE.

When Windmeijer (1992) used this data set he identified one outlier: this is a household which owns a new private car while it has a very low income and disposes of a business car. In the dataset this is observation 817. In the calculations reported in this book it has not been removed from the sample.

1.4 Notation

I have aimed at a consistent use of various scripts and fonts while respecting established usage, but the resulting notation is not altogether uniform. It is also often incomplete in the sense that it can only be understood in the context in which it is used. A full classification with a distinct notation for each type of expression and variable is so cumbersome that it would hinder understanding: as in all writing, completeness does not ensure clarity. I must therefore put my trust in the good sense of the reader. My main misgiving is that I found no room for a separate typographical distinction between random variables and their realization; in the end the distinction of boldface type was awarded to vectors and matrices as opposed to scalars.

The first broad distinction is that, as a rule, the Greek alphabet is used for unknown parameters and for other unobservables, such as disturbances, and Roman letters for everything else. The parameter β (and the vector $\boldsymbol{\beta}$) has the same connotation throughout, but λ and to a lesser extent α are used as needed and their meaning varies from one section to another. Greek letters are also occasionally employed for specific functions, like the normal distribution function.

In either alphabet there is a distinction between scalars and vectors or matrices. Scalars are usually designated by capital letters, without further distinction, but vectors and matrices are set in boldface, with lower-case letters for (column) vectors and capitals for matrices. I use a superscript ^{T} for transposition, the dash being exclusively reserved for differentiation.

The differentiation of vector functions of a scalar and of scalar functions of a vector habitually causes notational problems. In an expression like

$$\mathbf{y} = \mathbf{f}(X),$$

$\mathbf{y}$ is a column vector with elements that are functions of a scalar X. Differentiation will yield $\mathbf{f}'$, which is again a column vector. But in

$$Y = f(\mathbf{x}),$$

the scalar Y is a function of several arguments that have been arranged in the column vector $\mathbf{x}$. Differentiation with respect to (the elements of) $\mathbf{x}$ will yield a number of partial derivatives, which we arrange, by convention, in a row vector $\mathbf{f}'$. By the same logic, if $\mathbf{y}$ is an $r \times 1$ vector and $\mathbf{x}$ an $s \times 1$ vector, and if $\mathbf{y} = \mathbf{f}(\mathbf{x})$, $\mathbf{f}'$ is an $r \times s$ matrix of partial derivatives, and it should be named by a capital letter.

We use the standard terms of estimation theory and statistics, such as the expectation operator E, the variance of a scalar var and the variance-covariance matrix $\mathbf{V}$, applying them directly to the random variable to which they refer, as in EZ, var$Z$, and $\mathbf{V}\mathbf{z}$. The arguments on which these (and other) expressions depend are indicated in parentheses. Thus

$$\mathbf{Vz}\,(\boldsymbol{\theta})$$

indicates that the variance matrix of $\mathbf{z}$ is a function of the parameter vector $\boldsymbol{\theta}$, while

$$\mathbf{V}\hat{\boldsymbol{\theta}}$$

is the variance matrix of $\hat{\boldsymbol{\theta}}$. This is an estimate of $\boldsymbol{\theta}$ as indicated by the circumflex or hat above it. Again,

$$\hat{\mathbf{V}}\mathbf{z} = \mathbf{Vz}(\hat{\boldsymbol{\theta}})$$

indicates how an estimated variance matrix is obtained.

Probabilities abound. We write

$$\Pr(Y_i = 1)$$

for the probability of an event, described within brackets; the suffix i denotes a particular trial or observation. We will then continue as in

$$\Pr(Y_i = 1) = P_i = P(X_i)$$

where P_i is a number between 0 and 1 and $P(X_i)$ the same probability as a function of X_i. Its complement is denoted by

$$Q_i = 1 - P_i, \ \ Q(X_i) = 1 - P(X_i).$$

The vector $\mathbf{p}$ consists of a number of probabilities that usually sum to 1. At times we shall also make use of a different notation and use $\Pr(Y_i)$ for the probability of the observed value Y_i; in the binary case this is P_i if $Y_i = 1$ and Q_i if $Y_i = 0$, and it can be written as

$$\Pr(Y_i) = P_i^{Y_i} Q_i^{1-Y_i}.$$

Equations are numbered by chapter, but sparingly, and only if they are referred to elsewhere in the text.

2

The binary model

Binary discrete probability models describe the relation between one or more continuous determining variables and a single attribute. These simple models, probit and logit alike, account for a very large number of practical applications in a wide variety of disciplines, from the life sciences to marketing. In this chapter we discuss their background, their main properties, their justification and their use. Section 2.3 presents the latent variable regression model that is used as the standard derivation throughout this book. Although the emphasis is on the logit model, much of the discussion applies to the probit model as well.

2.1 The logit model for a single attribute

The logit model has evolved independently in various disciplines. One of its roots lies in the analysis of biological experiments, where it came in as an alternative to the probit model. If samples of insects are exposed to an insecticide at various levels of concentration, the proportion killed varies with the dosage. For a single animal this is an experiment with a determinate, continuously variable stimulus and an uncertain or random discrete response, viz. survival or death. The same scheme applies to patients who are given the same treatment with varying intensity, and who do or do not recover, or to consumer households with different income levels who respond to this incentive by owning or not owning a car, or by adopting or eschewing some other expensive habit. Married women may or may not take up paid employment and their choice is influenced by family circumstances and potential earnings; students' choices among options of further education are affected by their earlier performance. The class of phenomena or models thus loosely defined is variously referred to in the biological literature as *quantal variables* or as

9

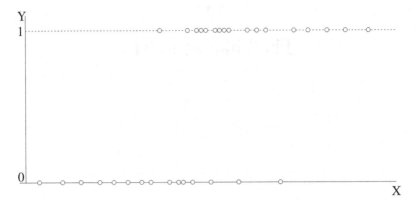

Fig. 2.1. Car ownership as a function of income in a sample of households.

stimulus and response models, in psychology and economics as *discrete choice*, and in econometrics as *qualitative* or *limited dependent variables*.

We examine the car ownership example more closely. The relation of household car ownership to household income can be observed in a household survey. The independent or determining variable is household income, which is continuous, and the dependent variable or *outcome* is ownership status, which is a discrete variable. For a single attribute (like car ownership as such) the outcome Y is a scalar which can take only two values, conventionally assigned the values 0 and 1. The event $Y = 1$ is habitually designated as a *success* of the experiment, and $Y = 0$ as a *failure*, regardless of their nature. In the present case we have

$$Y_i = 1 \text{ if household } i \text{ owns a car,}$$
$$Y_i = 0 \text{ otherwise.}$$

When these values are plotted against income X_i for a sample of households we obtain the scatter diagram of Figure 2.1.

A regression line could be fitted to these data by the usual Ordinary Least Squares (OLS) technique, but the underlying model that makes sense of this exercise does not apply.† One may of course still *define* a linear relationship, and make it hold identically by the introduction of an additive disturbance ε_i, as in

$$Y_i = \alpha + \beta X_i + \varepsilon_i.$$

† There is no short-cut formula for the OLS regression of Y on X. If X were regressed on Y, however, the regression line would pass through the mean incomes of car-owners and of non-car-owners.

In order to restrict the Y_i to the observed values 0 and 1, however, special properties must be attributed to the disturbance ε_i; the simple properties that are the main appeal of the OLS model will not do.

Instead, the natural approach to the data of Figure 2.1 is to recognize that Y_i is a discrete random variable, and to make the *probability* of $Y_i = 1$, not the value of Y_i itself, a suitable function of the regressor X_i. This leads to a *probability model* which specifies the probability of the outcome as a function of the stimulus, as in

$$P_i = \Pr(Y_i = 1) = P(X_i, \boldsymbol{\theta}),$$
$$Q_i = \Pr(Y_i = 0) = 1 - P(X_i, \boldsymbol{\theta}) = Q(X_i, \boldsymbol{\theta}).$$

Recall that, as a matter of notation, $\Pr(A)$ is the probability of the event A, $P(\cdot)$ is a probability as a function of certain arguments, and P a shorthand notation for either; $Q(\cdot)$ and Q are the complements of $P(\cdot)$ and P. Here $P(X_i)$ is a probability that is a function of X_i; the vector $\boldsymbol{\theta}$ of parameters that govern its behaviour has been added for the sake of completeness. In the sequel we shall often turn to a simpler and less formal notation. – Since P is a probability, it is bounded between 0 and 1. In the present use this is an open interval, and the two bounds are never attained. Many formulae in the sequel break down if this convention is disregarded.

The regression equation may be briefly revived by specifying

$$P(X) = \alpha + \beta X,$$

which is the *linear probability model*. It suggests the estimation of α and β by a regression of the Y_i on the X_i, with suitable embellishments like a correction for heteroskedasticity; see Goldberger (1964, p. 250) or, for a fuller treatment and a comparison with statistical discrimination, Ladd (1966). Reasonable results are obtained if a linear regression is fitted to observed frequencies that remain within a limited range, well away from their bounds 0 and 1; see Agresti(1996, p. 85) for an example. But technical improvements and special cases do not remove the principal objection, which is that the linear specification does not respect the limited range from 0 to 1 which is imposed on probabilities. If we wish the probability to vary monotonically with X and yet remain within these bounds, we must look for an S-shaped or *sigmoid* curve which flattens out at either end so as to stay within these natural limits. There are innumerable transformations that meet this requirement; its great analytical simplicity recommends the *logistic function,*

$$P(X) = \exp(\alpha + \beta X)/\left[1 + \exp(\alpha + \beta X)\right],$$
$$Q(X) = 1 - P(X) = 1/\left[1 + \exp(\alpha + \beta X)\right]. \tag{2.1}$$

Throughout the present chapter this particular probability function $P(\cdot)$ is denoted by Pl (with l for *logit* or *logistic*). This function is defined by

$$Pl(Z) = \exp Z/(1 + \exp Z),$$

so that (2.1) can be written as

$$P(X) = Pl(\alpha + \beta X).$$

In the present section the argument Z is a linear function of a single regressor, but it can equally well be a function of several regressors $\mathbf{x}^T \boldsymbol{\beta}$.

There is no direct intuitive justification for the use of the logistic function; we return to this point in the next section. Here we shall first examine its properties. It follows from the definition of $Pl(Z)$ that

$$1 - Pl(Z) = Pl(-Z), \tag{2.2}$$

so that the pair of (2.1) can be rewritten as

$$P(X) = Pl(\alpha + \beta X),$$
$$Q(X) = Pl(-\alpha - \beta X).$$

The behaviour of the logistic function is simple. It follows the sigmoid curve shown in Figure 2.2, with a point of inflection at $Pl(0) = 0.5$, that

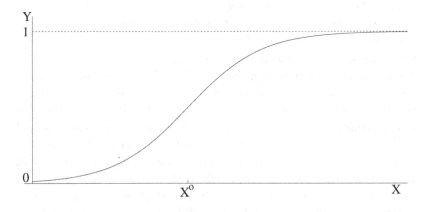

Fig. 2.2. The logistic curve $Pl(\alpha + \beta X)$.

is at X° which satisfies $\alpha + \beta X^\circ = 0$, or $X^\circ = -\alpha/\beta$. It follows from (2.2) that the curve is symmetrical around this midpoint, in the sense that the same curve is obtained if we reverse the direction of the X-axis and then turn the diagram upside down. The original curve for $Pl(X)$ and this mirror image for $Ql(X)$ cross at the midpoint X° with $Pl(0) = Ql(0) = 0.5$. The *slope* of the curve is governed by β, while its *position* with respect to the X-axis is set by the location parameter or intercept α. For continuous X, the effect of (a change of) X on $Pr(Y = 1)$ is given by the derivative of the curve, and this varies quite strongly with the level of P and therefore with the point at which it is evaluated. This derivative is

$$Pl'(\alpha + \beta X) = Pl(\alpha + \beta X)\left[1 - Pl(\alpha + \beta X)\right]\beta$$
$$= P(X)Q(X)\beta. \tag{2.3}$$

It goes to 0 as P approaches its bounds of 0 or 1, and has a maximum of $\beta/4$ at the midpoint. All we can say is that its sign is determined by the sign of β. For a positive β, $Pl(\alpha+\beta X)$ increases monotonically from 0 to 1 as X ranges over the entire real line, which is precisely what is required.

In some fields, like epidemiology, the analysis traditionally bears on the *odds* rather than on the probability of a particular event; this is defined as

$$\text{odds } P(Z) = P(Z)/[1 - P(Z)]. \tag{2.4}$$

While probabilities are confined to the interval $(0, 1)$, odds range over the positive half of the real number axis. Upon taking logarithms we obtain the *log odds* or *logit* of a probability (or frequency),

$$\log \text{odds} = \text{logit}\,[P(Z)] = R(Z) = \log\left\{P(Z)/[1 - P(Z)]\right\}. \tag{2.5}$$

Once more we have two names for the same quantity, which may vary over the entire real number axis from $-\infty$ to $+\infty$. For logistic probabilities the odds are

$$\text{odds}\,(Z) = \exp(Z) = \exp(\alpha + \beta X),$$

and the log odds or logit is

$$\text{logit}\,[Pl(Z)] = Rl(Z) = \log\left\{Pl(Z)/[1 - Pl(Z)]\right\} = Z,$$

or

$$\text{logit}\,[Pl(\alpha + \beta X)] = \alpha + \beta X. \tag{2.6}$$

The simplicity of this inverse transformation is a major attraction of the logistic model. The linear form of the logit is just as much an identifying characteristic of the model as the original logistic function of (2.1) for the probability; the two are equivalent. Since the probability $P(Y_i = 1)$ is equal to the expected value of the random variable Y_i, the logit transformation can also be regarded as a *link function* from the theory of General Linear Models, giving $E(Y_i)$ as a linear function of X_i; see McCullagh and Nelder (1989, p. 31).

These arguments suggest that the model may well be derived from the other end by starting off from a consideration of odds and log odds, and then arriving at the logistic function. This is common usage in case–control studies, which deal with the special case of a single binary categorical regressor variable X. The log odds therefore takes two values only,

$$\log \text{odds } (P) = \alpha + \beta \text{ for } X_i = 1, \quad \log \text{odds } (P) = \alpha \text{ for } X_i = 0.$$

As a result, the log of the *odds ratio* is simply equal to β, and this is a measure of the effect of X. This is the central approach of case–control studies, with the odds determined directly from sample frequencies; we return to the subject in Section 6.4.

Several summary characteristics of the logistic curve for a continuous X can be derived from the two parameters α and β. Insecticides and similar products are often graded by the midpoint concentration level $X° = -\alpha/\beta$, known as the *50% effective dosage*, or *ED50*. But in many analyses the main interest is not in the level of P but in the effect of particular regressor variables on P. This is reflected by the derivative (2.3), evaluated at a suitable value of X or P. In economic analyses, such effects are habitually expressed in *elasticities* of the form

$$\partial \log W / \partial \log V,$$

for any pair of causally related variables W and V. The popular interpretation is that this represents 'the percentage change in W upon a 1% change in V', and the major reason why it is preferred to the derivative is that it is invariant to the arbitrary units of measurements of both variables. In the present case, however, the dependent variable is a probability, and its scale is not arbitrary since it ranges from 0 to 1. We therefore recommend the use of *quasi-elasticities*, defined as

$$\eta = \partial P(X) / \partial \log X,$$

and given in the present instance by

$$\eta(X) = Pl(\alpha + \beta X)[1 - Pl(\alpha + \beta X)]\beta X. \tag{2.7}$$

This indicates the percentage point change of the probability upon a 1% increase of X. Like the derivative (and the elasticity) its value varies with P and hence with X. These measures are therefore usually evaluated at the sample mean or some other convenient point. Note that the probability at the sample mean $\bar{X}$, $Pl(\alpha + \beta\bar{X})$, is not in general equal to the sample mean of Y (the sample frequency of $Y = 1$), because of the nonlinearity of $P(\cdot)$. As P and Q sum to 1, their derivatives sum to 0, and so do the quasi-elasticities which are equal but of opposite sign. Ordinary elasticities do not have this property.

This is not the only solution to the problem that the value of β (and hence the derivative (2.3)) varies with the arbitrary scale of X. A simple alternative from sociological studies is to switch to *standardized* regressors: X is scaled by its standard deviation, or if it is left unscaled β, and the derivative, are multiplied by the same factor. This renders the effects of different regressors comparable.

In the controlled laboratory conditions of bio-assay, the concentration of the insecticide is the only observable cause of death, all other conditions being kept constant as far as is feasible. Epidemiological studies often concentrate on a single treatment variable, and also seek to limit heterogeneity in other respects; still a case can be made for taking into account other sources of variation, insofar as the observations permit. Car ownership is of course affected by many household characteristics other than income, and in the design of household surveys their variation cannot be avoided. When studying a random sample from the population we must therefore control for several obvious additional determinants of car ownership, like family size and the degree of urbanization, by including them in the analysis. The logit model easily accommodates additional variables, as in $P(\mathbf{x_i}) = Pl(\mathbf{x_i}^T\beta)$, where $\mathbf{x}$ and β are vectors. Any linear function of relevant regressor variables can thus be inserted in the logistic function; its argument may be treated exactly like a linear regression equation. The vector $\mathbf{x}$ invariably includes a constant '1' with the intercept parameter β_0 as its coefficient. Just as in a regression function the other regressor variables may be transformed, e.g. by taking logarithms or by adding squares or products ('interaction terms'). Particularly rich and flexible nonlinear scale variations of the regressor variables are provided by the use of fractional polynomials, discussed at some length by Hosmer and Lemeshow (2000, Ch. 4). Categorical regressor variables can be represented by one or more $(0, 1)$ dummy variables.

So far we have usually designated the independent variables, determinants or explanatory variables of the model as regressors, following the usage in economics and econometrics; in the biomedical and statistical literature they are known as covariates. From now on we shall use these two terms without discrimination.

2.2 Justification of the model

Even when techniques like ordinary regression or fitting a simple logit are almost mechanically applied, the original arguments leading up to the model and establishing its properties are still present in some distant recess of the analyst's mind and affect the understanding of the findings. The logit model has no immediate intuitive appeal, but it can be justified in several ways: as a simple approximation, by the consideration of random processes, or from models of individual behaviour. Some of the latter permit an interpretation of economic choices in terms of utility. The various arguments are here discussed for a single regressor variable X, but they are equally valid for models with several covariates.

The approximation argument is quite similar to viewing the linear regression equation as an approximation to a more complex analytical relation between the regressors and the dependent variable. The logit model can be regarded in the same light as an approximation to any other probability model, provided the log odds is taken as the starting point; in biomedical research its primacy is generally accepted. By (2.5) this is defined for any $P(X)$ as

$$R(X) = \log \{P(X)/[1 - P(X)]\}.$$

A Taylor series expansion around X° yields

$$R(X) = R(X^\circ) + R'(X^\circ)(X - X^\circ) + \text{remainder}$$
$$= [R(X^\circ) - R'(X^\circ)X^\circ] + R'(X^\circ)X + \text{remainder}.$$

The first term is a constant, the second is linear in X, and the remainder represents terms in the higher-order derivatives. If $P(X)$ is the logistic function the linear function holds exactly and the remainder is zero, as in (2.6). For other probability functions the linear part constitutes an approximation. Its quality depends on the form of $P(X)$, and in particular on its higher derivatives; but these do not easily lend themselves to further discussion.

The logit model can also be obtained from a random process in which individuals alternate between two states, such as sickness and health, employment and unemployment. The durations or *spells* of either state are nonnegative random variables; if they depend on a regressor X, various models will lead to expected durations of the intervals spent in states 0 and 1 of the form $\exp(\alpha_0 + \beta_0 X)$, $\exp(\alpha_1 + \beta_1 X)$. Under quite general conditions the probability of finding an individual drawn at random in state 1 is then

$$P(X) = \exp(\alpha_1 + \beta_1 X)/[\exp(\alpha_0 + \beta_0 X) + \exp(\alpha_1 + \beta_1 X)];$$

see Ross (1977, Ch. 5). Since the numerator and the denominator can be multiplied by $\exp(\delta + \zeta X)$ without affecting $P(X)$, the original parameters cannot be separately established, and some simplification is in order; upon writing $\alpha = \alpha_1 - \alpha_0$, $\beta = \beta_1 - \beta_0$ the logit model (2.1) is obtained.

There is a wide range of models of individual behaviour involving random elements that lead to a probability model which turns into the logit model upon further specification of the random term. This is the class of *threshold* models, so named after their classic use in bio-assay or the testing of insecticides. The *stimulus* X is the dosage of an insecticide, a nonrandom variable set by the analyst; the discrete *response* is death of the insect, determined by a comparison of X with its *threshold value* or *tolerance level* ε. The response function for a single experiment looks like the step function of Figure 2.3; for a given insect the threshold is a constant. It is treated as a random variable because it varies among

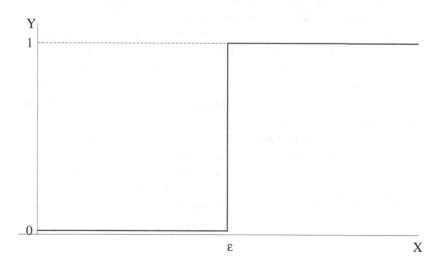

Fig. 2.3. An individual response function in the threshold model.

insects and we are considering an individual drawn at random from this population. For observation i we have

$$\Pr(Y_i = 1) = P(X_i) = \Pr(\varepsilon_i \leq X_i)$$

so that

$$\Pr(Y_i = 1) = F(X_i)$$

where F is the distribution function of ε. In the classic example this is a normal distribution, and it is transformed to the *standard* normal distribution Φ as in

$$P(X_i) = \Phi\left(\frac{X_i - \mu}{\sigma}\right) = \Phi(\alpha + \beta X_i).$$

Other assumptions are needed to obtain a logistic function; we deal with this in the next section.

This threshold model for the responses of living organisms is applicable to a wide range of human behaviour. If examinations take the form of multiple-choice questionnaires success consists in giving the correct answer, and this depend on the students' ability surpassing a certain threshold; consumer expenditure on particular items, household ownership of durable goods or housing decisions of a family can all be regarded as responses to the stimuli of income, prices or advertising campaigns. According to the circumstances the model may be elaborated by introducing specific determinants of the stimulus and of the threshold, and by speculating about the nature of the random elements involved. In the above example, the tolerance level of the subject insect is random and the stimulus is not, but in other cases it may be the other way around, and in still other cases it may be sensible to introduce several distinct stochastic elements. In the multiple-choice example, the students' ability is not directly observable, but a function of a number of characteristics, while the threshold depends on the difficulty of the set question; both may contain a random term. In a simple *utility* or *discrete choice* model of the choice between two options (like owning a car or not) the individual attaches separate random utilities to the two possible states $Y = 1$ and $Y = 0$. Both utilities vary with the same covariate, as in the pair

$$U_1 = \alpha_1 + \beta_1 X + \varepsilon_1,$$
$$U_0 = \alpha_2 + \beta_2 X + \varepsilon_2.$$

Utility maximization implies that the state with the higher utility obtains,

or

$$P(X) = \Pr(Y = 1) = \Pr(U_1 > U_0)$$
$$= \Pr[(\alpha_1 - \alpha_2) + (\beta_1 - \beta_2)X + (\varepsilon_1 - \varepsilon_2) > 0].$$

The parameters α reflect utility levels of the two alternatives, and the β the effect of the regressor on these utilities; the utility differential is the stimulus that triggers the response, with a threshold of zero. Once more, however, the parameters of the initial formulation cannot be established, as they can be varied without affecting the probability of the outcome as long as utility differences remain the same. Once more we are compelled to reduce the model to a single inequality of a linear function of the regressor and a (composite) random term. As in the insecticide example the behaviour of P is governed by the latter's distribution, which must still be specified to complete the model. The reformulation in utility differentials, not levels, is in keeping with the view that utility measures at best permit the ordering of alternatives. This is a very simple model of utility maximization, if only by its limitation to the comparison of two alternatives; a much more sophisticated model for any number of alternatives will be presented in Section 8.2.

In the utility example we have been obliged to simplify the model from its first, rather grandiose inception because we have invented more than we can observe. In economics this is a common occurrence. The natural description of the process under consideration leads to a surfeit of parameters which cannot be established from the observed facts since there is no one-to-one correspondence from parameter values to the outcome or its probability distribution. Even if this probability distribution were known there is no unique vector of parameter values associated with it; several parameter vectors are observationally equivalent in the sense that no observed sample, however accurate or large, permits us to discriminate between them. In economics this is known as lack of *identification* of the model. In the utility model the issue is resolved by replacing the four parameters α_1, α_2 and β_1, β_2 by two pairwise differences (with similar simplifications of the disturbances). The original parameters are *structural* parameters, in the sense that they correspond to behavioural constants that are the same across observations from different sources and different contexts; in other models, like variations of the insecticide example, they may represent physiological constants. If these structural parameters are unidentified, the solution is to reduce their number by fixing *a priori* values for some parameters, or by replacing several structural parameters by a smaller number of functions, like the differences in the utility model. The resulting composite parameters that can be established from data are known as *reduced form* coefficients. We shall presently give another instance of underidentification and its resolution in this manner.

The somewhat romantic flavour of these behavioural models is enhanced by further details. In many instances the stimulus variable like income or the dosage of a poison is always positive and has a skew distribution, and $\log X$ often gives better results than X in terms of fit; the logarithmic transformation is standard practice in the original field of bio-assay. This is sometimes attributed to the *law of proportionate effect* or *Weber–Fechner law*, which says that the physiological impact of a stimulus is an exponential function of its physical strength. This law was empirically established in the middle of the 19th century by such experiments as asking subjects to select the heavier of two weights; the standard reference is Fechner (1860).

All these models assume a causal relation between the original stimulus X, perhaps some intermediate latent variable like the impact or utility, and the outcome Y. In the 'approximation' and 'alternating states' arguments, X moreover clearly stands for external determinants. There is always a definite asymmetry between X and Y with X the cause and Y the (uncertain) effect, without any feedback; X is not affected by the actual outcome Y. In Section 3.2 this basic assumption is used to justify the treatment of X in estimation. Any idea of a joint distribution of Y and X without a clear causal link from the one to the other is alien to the models under review, and will not be followed up in this book, apart from a brief discussion of discriminant analysis in Section 6.1.

The threshold model and other behavioural models all end up in a probability model whereby the outcome depends on a linear inequality in the regressors and a random disturbance. Further assumptions are needed to turn this into logistic regression, and this is the subject of the next section.

2.3 The latent regression equation; probit and logit

The statistical specification of various behavioural models consists of a linear regression equation for a continuous *latent* variable Y^* and an inequality that establishes the observed discrete variable Y. This latent regression model will be invoked on several occasions throughout this book, and we shall here present it at some length.

The model consists of two equations. The first is the latent variable regression equation

$$Y_i^* = \mathbf{x}_i^T \boldsymbol{\beta}^* + \varepsilon_i^* \tag{2.8}$$

where the vector $\mathbf{x}_i$ invariably includes a unit constant and $\boldsymbol{\beta}^*$ an intercept. This has all the properties of a classic linear regression equation. The regressors or covariates $\mathbf{x}_i$ are known constants, ε^* is a random disturbance that is uncorrelated with the regressors, and $\boldsymbol{\beta}^*$ represents unknown parameters of the underlying process that determines Y^*. The *latent variable* Y_i^* is not observed, but the $(0, 1)$ outcome Y_i is; these observations are determined by the inequality condition

$$Y_i = 1 \ \text{if} \ Y_i^* > 0,$$
$$Y_i = 0 \ \text{otherwise.} \tag{2.9}$$

It follows at once that

$$\begin{aligned} \Pr(Y_i = 1) = P(\mathbf{x}_i) &= P(\varepsilon_i > -\mathbf{x}_i^T \boldsymbol{\beta}^*) \\ &= 1 - F(-\mathbf{x}_i^T \boldsymbol{\beta}^*), \end{aligned} \tag{2.10}$$

where $F(\cdot)$ is the distribution function of ε^*.

The function $P(\mathbf{x}_i)$ now depends entirely on the distribution of ε^*. We make a number of assumptions about this distribution. The first is that it has *zero mean*,

$$\mathrm{E}\varepsilon^* = 0.$$

This is a universal assumption, here as well as in the case of ordinary regression; but it is not a *natural* assumption, for there is no reason why ε^* (which covers, among other things, the effect of missing regressor variables) should not have a nonzero mean μ^*. But this mean would not be identified, since it is confounded with the intercept β_0^* of the regression equation. By assuming that the mean is zero any nonzero μ^* is absorbed into the intercept, and this may account for the general lack of interest in this parameter.

In later chapters we shall come across other instances of irrelevant factors or nuisance parameters that enter into the intercept but not into the other elements of $\boldsymbol{\beta}$. If the need arises to emphasize this distinction we shall partition $\mathbf{x}$ and $\boldsymbol{\beta}^*$ or other similar coefficient vectors as

$$\mathbf{x} = \begin{bmatrix} 1 \\ \tilde{\mathbf{x}} \end{bmatrix},$$

$$\boldsymbol{\beta}^* = \begin{bmatrix} \beta_0 \\ \tilde{\boldsymbol{\beta}}^* \end{bmatrix}. \tag{2.11}$$

$\mathbf{x}$ thus consists of a unit constant and the elements of $\tilde{\mathbf{x}}$, which are termed *proper* regressors or covariates, and $\boldsymbol{\beta}^*$ of the intercept and the *slope coefficients* of $\tilde{\boldsymbol{\beta}}^*$.

The second assumption is that the distribution of ε^* is *symmetrical* around zero. As a result (2.10) can be rewritten as

$$\Pr(Y_i = 1) = P(\mathbf{x}_i) = F(\mathbf{x}_i^T \boldsymbol{\beta}^*). \tag{2.12}$$

Thus the sigmoid curve of Section 2.1 is a cumulative distribution function.

The third assumption fixes the variance of ε^*. Like the assumption of a zero mean this is needed to ensure identification. The need arises because the inequality (2.9) is invariant to changes of scale of all its terms. This indeterminacy is resolved by imposing a set value c^2 on the variance of the disturbances, putting their standard deviation σ^* equal to c. Both sides of (2.8) are then multiplied by c/σ^*, so that it is replaced by

$$Y_i^\circ = \mathbf{x}_i^T \boldsymbol{\beta} + \varepsilon_i$$

with

$$Y_i^\circ = Y_i^* \frac{c}{\sigma^*}, \quad \boldsymbol{\beta} = \boldsymbol{\beta}^* \frac{c}{\sigma^*}, \quad \varepsilon_i = \varepsilon_i^* \frac{c}{\sigma^*}. \tag{2.13}$$

and Y_i is determined as before by the sign of Y_i°. Note in particular the change to a new set of *normalized* or *standardized* regression coefficients $\boldsymbol{\beta}$. This normalization of $\boldsymbol{\beta}$ is usually passed by as a merely technical adjustment, but it may have material consequences. It is motivated by the need to ensure identification, and it is certain that in the terms of the earlier discussion the elements of $\boldsymbol{\beta}$ are reduced form coefficients. It is not so certain however that the $\boldsymbol{\beta}^*$ represent structural coefficients, for they themselves may have been earlier derived from a more elaborate underlying model, as in the utility maximization example of the preceding section.

We return to the choice of the distribution function of (2.12). As all distribution functions produce sigmoid curves that rise from 0 to 1, *any* symmetrical distribution with zero mean will do; but in practice only two distributions are chosen, the normal and the logistic distribution.

The *normal distribution* is the natural specification of any residual or otherwise unknown random variable; it has a direct intuitive appeal, if only because of its familiarity. Here we assume a *standard* normal distribution with zero mean and unit variance. Its density is

$$\phi(Z) = \frac{1}{\sqrt{(2\pi)}} \exp\left(-\frac{1}{2}Z^2\right),$$

and the distribution function is

$$\Phi(Z) = \frac{1}{\sqrt{(2\pi)}} \int_{-\infty}^{Z} \exp\left(-\frac{1}{2}t^2\right) dt.$$

If this is used as a probability function it is denoted by $Pn(Z)$, just as $Pl(Z)$ denotes the logistic. This specification

$$\Pr(Y_i = 1) = P(\mathbf{x}_i) = Pn(\boldsymbol{\beta}^T \mathbf{x}_i)$$
$$= \Phi(\boldsymbol{\beta}^T \mathbf{x}_i) \qquad (2.14)$$

is the *probit model.* Historically the probit model precedes the logit, as is recounted in Chapter 9. In addition to zero mean the standard normal distribution has unit variance: the scaling constant c introduced above is equal to 1. There is no reason to suppose that the standard deviation of ε^* has this particular value, and the assumption is solely prompted by the need for identification.

The major merit of the *logistic density* for ε^* is that it leads to the logit model; we may as well derive it from the desired distribution function, which is

$$F(Z) = Pl(Z) = \exp Z/(1 + \exp Z).$$

This gives the density

$$f(Z) = \exp Z/(1 + \exp Z)^2. \qquad (2.15)$$

This is known as the logistic density, the sech^2 or the Fisk density; see Johnson and Kotz (1970, vol. 2, Ch. 22) for details. I know of no experiment or process which engenders this distribution in a natural way. The density has mean zero and variance $\lambda^2 = \pi^2/3$, or standard deviation

$$\lambda = \pi/\sqrt{3} \approx 1.814.$$

The *standardized* logistic distribution with zero mean and unit variance therefore has the distribution function

$$F(Z) = \exp \lambda Z / (1 + \exp \lambda Z)$$

and density

$$f(Z) = \lambda \exp \lambda Z / (1 + \exp \lambda Z)^2 \, .$$

If we proceed in the reverse direction, from the regression equation via the density to the distribution, the parameter vector $\boldsymbol{\beta}^*$ of (2.8) must again be normalized in the interest of identification; but now the set value c of the standard deviation of the disturbances must be λ, not 1. With the same parameters $\boldsymbol{\beta}^*$ of the original regression equation, and the same residual variance σ^{*2} of ε^*, the coefficients of the logit model are thus a factor λ greater than the coefficients of the probit model.

While the probit specification is analytically less tractable than the logistic function, the two functions are quite similar in shape, as was first demonstrated by Winsor (1932). This can be illustrated graphically and numerically.

Figure 2.4 shows the two densities in standardized form with zero mean and unit variance, as given above. The logistic density has a higher peak than the normal, it is thinner in the waist, and it has thicker

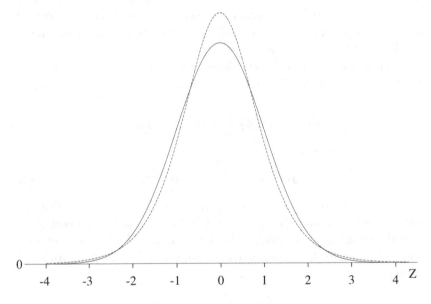

Fig. 2.4. Logistic (broken line) and normal (solid line) density functions with zero mean and unit variance.

tails, but these appear only at fairly extreme values of more than three standard deviations from the mean. This is measured by the *kurtosis* of the distribution, which is 1.2 for the logistic distribution as against zero for the normal.

For the course of probabilities as a function of X in the two models we should consider distribution functions rather than densities. The distribution function of the logistic was shown in Figure 2.2; it might be compared in the same diagram with a normal distribution function, drawn to the same scale, but the two curves would be virtually indistinguishable. A numerical comparison is therefore in order. In panel A of Table 2.1 the two standardized distribution functions given earlier and their difference Δ are tabulated for deviations from the mean. This shows how the two probabilities vary with the argument, adjusted to a common scale. Over the range of P from 0.1 to 0.9 the logit rises more steeply than the normal, but beyond these values the position is reversed. Panel B of the table gives a tabulation of the inverse functions and their ratio. This shows that after standardization to the common

Table 2.1. *Comparison of logit and probit probabilities.*

	A				B		
X	$F(X)$	$\Phi(X)$	Δ	P	logit	probit	ratio
0.00	0.500	0.500	0	0.50	0	0	-
0.10	0.545	0.540	0.005	0.51	0.022	0.025	1.136
0.20	0.590	0.579	0.011	0.52	0.044	0.050	1.138
0.30	0.633	0.618	0.015	0.53	0.066	0.075	1.138
0.40	0.674	0.655	0.019	0.54	0.088	0.100	1.136
0.50	0.712	0.692	0.020	0.55	0.111	0.126	1.136
0.75	0.796	0.773	0.023	0.60	0.224	0.253	1.133
1.00	0.860	0.841	0.019	0.65	0.341	0.385	1.129
1.25	0.906	0.894	0.012	0.70	0.467	0.524	1.123
1.50	0.938	0.933	0.005	0.75	0.606	0.675	1.114
1.75	0.960	0.960	0.000	0.80	0.764	0.842	1.101
2.00	0.974	0.977	−0.003	0.85	0.956	1.036	1.084
2.25	0.983	0.988	−0.005	0.90	1.211	1.282	1.058
2.50	0.989	0.994	−0.005	0.95	1.623	1.645	1.013

Part A: for negative Z, take complements.
Part B: for complements of P, reverse signs.
Normal values from Fisher and Yates (1938).

variance of 1 (which implies a reduction of the logit coefficients by a factor λ), the logit is systematically too small in absolute value, at least for moderate values of P. To bring about a closer correspondence with the probit over this range the standardized logit coefficients must be raised by some 10%; in other words, the original coefficients should be reduced by a factor of about 1.6 rather than $\lambda = 1.814$ to match the probit coefficients. This is in line with the conclusion of Amemiya (1981), who equates the slope of the two curves at the midpoint of $P = 0.5$, $X = 0$ and finds a ratio of 1.6. In many empirical applications the midpoint is however by no means representative of the observed sample. Sample proportions of 0.8/0.2 and beyond are more common than 0.5/0.5, and the range of the probabilities of the sample observations will then lie far outside the limits of 0.1 and 0.9. This may lead to different ratios of the logit to probit coefficients if both curves are fitted to the same data.

The conclusion is that by judicious adjustment of their coefficients logit and probit models can be made to virtually coincide over a fairly wide range. The ratio of the two sets of coefficients depends on the actual range of probabilities which the two curves must describe; it varies from one application to another. Upon fitting the two curves to the same data the ratio of the estimated coefficients may lie anywhere between 1.6 and 2.0. Nevertheless, the two probability functions are as a rule virtually indistinguishable, and it is practically impossible to choose between them on empirical grounds. As for theoretical arguments, they have most force for multinomial models; see Section 8.4.

2.4 Applications

The primary product of logit and probit routines of program packages consists of estimates of the parameters β and of their variances. This is enough for the investigation of statistical association, but not for two equally important purposes of empirical research, namely *selection* or *discrimination* and *(conditional) prediction*. We briefly discuss these three forms of application, and quote some typical examples to convey the flavour of applied work; but logit and probit analyses are used as widely as ordinary regression, and it would be pointless to attempt a survey of their subject matter.

Statistically significant or just convincing values of slope coefficients or of odds ratios establish a *statistical association* between covariates and the outcome under consideration. In many cases this is all the analyst is looking for; such findings may lead the way to further substantive research, or they may just satisfy scientific curiosity. Gilliatt (1947) performed laboratory experiments to establish vasoconstriction of the skin, which means that some people get white knuckles when they take a deep breath quickly. The data set of 39 observations is a favourite illustration in Finney's early treatise on the probit model (Finney (1971), first published in 1947). Another illustration which found its way into a textbook is the study of depression in Los Angeles by Frerichs et al. (1981). These authors have screened over 30 potential explanatory variables in a survey among a thousand people or so; part of the data found their way into the textbook of Afifi and Clark (1990) and from there into the BMDP manual of 1992. The well known textbook of Hosmer and Lemeshow (2000) illustrates logistic regression for the incidence of coronary heart disease and of low weight at birth and for the survival of patients in intensive care. The latter example is taken from an original study by Lemeshow et al. (1988) which employs hospital records for about 800 patients to assess the effect of treatment variables like the characteristics of intensive care units after controlling for the conditions of the patients. In the same vein, Silber et al. (1995) analyse over 70 000 hospital admissions for simple surgery, employing about 50 patient covariates (including interactions) and 12 hospital characteristics. The importance of such analyses for those who have the misfortune to need surgery or intensive care is self-evident. Returning to textbook examples we cite Agresti (1996), who demonstrates logistic regression (and other techniques) on data for Florida horseshoe crabs. The nesting females of this species have one or more male crabs attached to them, who are called satellites; the presence (and number) of these satellites is related to such attractive traits of the female as her colour, spine condition, weight and carapace width. An example of settling a factual issue in economics is the analysis of Oosterbeek (2000), who employs a probit analysis to resolve the public policy issue of adverse selection, that is the question whether or not individuals are more inclined to take out additional health insurance if their personal risk profile is unfavourable (they are). And Layton and Katsuura (2001) employ logit and probit models to identify turning points in the business cycle (this fails). In epidemiology there is an extensive literature of case–control studies that establish what conditions contribute to particular diseases, as a preliminary to closer physiological examination; the relationship of cigarette smoking and lung cancer is the classic example from an almost endless list.

The second application of discrete probability models is in *selection* or *discrimination*, and on this subject we offer more arguments and fewer anecdotes. The estimates of β serve to calculate predicted probabilities for individuals or items with given covariates, and these probabilities are then used for their classification, identification or segmentation into two (or more) groups. This use of discrete probability models is an alternative to *discriminant analysis*, which deals with the same classification problem in a more direct manner (see Section 6.1). Examples of target groups are individuals who are prone to a certain disease, and hence eligible for preventive treatment; prospective customers who are interested in a particular product; potential borrowers who are likely to default; firms heading for bankruptcy, and corporations that are likely objects of take-over bids. In all such dichotomies the probabilities are calculated for a group of subjects, and the individuals are ranked accordingly; it remains to set a critical *cut-off* value that separates the target group from the others.

This choice is of more than academic interest when selection has direct consequences for the selecting agent and for the chosen subjects. In marketing and financial choices it is a matter of maximizing the agent's money profit, or rather of minimizing the expected money loss, since the actual decision is always suboptimal. This is an elementary application of *decision theory*. There are two types of misclassification and we assume their costs are known: C_1 is the cost of erroneously attributing a successful outcome to an individual, and C_0 the cost of the reverse error. The cut-off criterion P° is applied to the (estimated) probabilities with a distribution $F(P)$. The expected loss due to misclassification is then

$$\mathrm{E}(L) = C_1 \int_{P^\circ}^1 (1-P)\mathrm{d}F(P) + C_0 \int_0^{P^\circ} P\mathrm{d}F(P),$$

the minimum condition is

$$-C_1(1 - P^\circ) + C_0 P^\circ = 0,$$

and the optimal value of the cut-off criterion is

$$P^\circ = \frac{C_1}{C_1 + C_0}.$$

At equal costs the criterion is 0.5; this is a popular (but quite unfounded) choice in situations of ignorance. In most applications the costs are highly unequal: upon comparing the profit on a sale with the cost of a circular letter, or the cost of default or bankruptcy with the profit on a loan, we arrive at cut-off rates of a few per cent. But then the proportions of sales or of defaults in a random sample of clients or borrowers are of the same order of magnitude, and so are the estimated probabilities.† The

---

† This is not as obvious as it may seem; see Cramer (1999).

dilemmas of dealing with default risk in real life are more complex, even if the consequences of various courses of action are easily expressed in money terms: for excellent demonstrations we refer to the identification of take-over targets by Palepu (1986) and to the analysis of the default of credit card holders by Greene (1992).

Binary models are also used to describe *sample selection* in a number of econometric models. In the *Tobit model*, due to Tobin (1958), household expenditure on durable goods Y^* is described by a linear regression on income and age. Negative values of Y_i^*, indicating that the household would rather undo its purchases of durable goods, are recorded as zero. This is a special case of the model of Section 2.3: (2.8) is not completed by (2.9) but by

$$Y_i = Y_i^* \text{ if } Y_i^* > 0,$$
$$Y_i = 0 \text{ otherwise.}$$

With a normal distribution of the disturbances, the proportion of zero expenditure at given covariates follows a probit function (with a negative slope), while the nonzero observations have a *censored* normal distribution; both are governed by the same regression equation. In the more general model of Heckman (1979), the selection of the observations is a separate mechanism, distinct from the main relationship under review. If this is the relation of women's labour supply (hours worked) to wages, there is a separate discrete model with a latent regression that determines whether women work at all. The two equations are related through the correlation of their disturbances, which must be taken into account in their estimation. In both types of model, Tobin's and Heckman's, the binary model (invariably a probit) plays only an auxiliary role. For further particulars the reader is referred to Maddala (1983) or Wooldridge (2002).

The third use of the results of empirical analyses is *(conditional) prediction*. This is a vast but uncharted field which is badly in need of a little theoretical discipline. Its more complex forms, often based on multinomial extensions of the model embracing more than two alternatives, are the mainstay of applied economics like marketing and policy studies. Discrete models have flourished in transport studies, and the sophisticated theories of Chapter 8 have largely been developed in practical studies in preparation of public transport systems like San Francisco's BART.

At the simplest level, economists wish to know the effect of (a change of) a particular regressor on the outcome. The standard practice is to consider the expected value of this random variable, and in ordinary regression the slope coefficient gives the derivative of this expectation with respect to the regressor concerned. In discrete models, we may likewise assess the effect of (changes in) covariates on the probability of success by the derivatives, elasticities and quasi-elasticities of Section 2.1. Their value depends strongly on the point at which they are evaluated, and it is uncertain whether the choice of the sample mean or some other representative point will indeed yield an estimate that is applicable to the sample or to the population from which it is drawn. Yet this is what is required, for interest centres on the effect of adjusting a regressor on the incidence of outcomes in an aggregate and not on individual outcomes.

This is fortunate, for the practice of predicting expected values makes little sense for a single discrete outcome Y_i. At best we have a precise estimate of the probability $P(\mathbf{x}_i)$ which is the expected value of Y_i; but a probability is not an admissible predictor of an outcome that can only take the values 0 or 1. It is a basic requirement of a prediction that it is of the same dimension as its object; the temperature of sea water must be predicted in degrees, and the balance of payments as a sum of money, and it is of no use to predict the one as green and the other as blue. The probability must therefore be transformed into a discrete $(0, 1)$ variable. In the literature this is often resolved by introducing a *prediction rule* to the effect that the outcome with the larger probability is adopted; in the binary case this is equivalent to generating predicted values by classification with a cut-off point of 0.5. There is no basis for this procedure; its repeated application to the same given covariates will not lead to an acceptable result, nor will its use in a heterogeneous group of individuals. The correct solution is to acknowledge the random character of the outcome and to assign the value of 1 to Y_i at random with the estimated probability $\hat{P}_i$, putting it otherwise at 0. But this method will only work if the individual prediction is repeated, or if the procedure is applied to a group of individuals to yield a prediction of the aggregate frequency; it offers no satisfactory solution to the problem of predicting a *single* individual outcome. We give up on that problem; it is insoluble, and it is of no practical interest. In practice the aim of empirical work in forecasts and policy analyses is the conditional prediction of the aggregate incidence of success in (sub)populations, and, by varying the conditions, the assessment of the effect of changes in the covariates on the result. Examples are the future course of car ownership under certain demographic and economic developments, the future composition of manpower by schooling, the demand for urban public transport under a new fare structure, or, on a smaller scale, the number of parking lots in a housing estate, the number of passengers that will turn up for for a particular flight, or the number of patients of a given hospital that may need intensive care treatment.

In all these cases it is supposed that we know some characteristics of N individuals that constitute the *prediction sample*, and that we have estimated probabilities $\hat{P}_i(\mathbf{x}_i)$ as a function of these characteristics. The expected frequency of success (variously defined) in the group is then predicted by the mean probability

$$\bar{P} = \frac{1}{N} \sum_i \hat{P}(\mathbf{x}_i).$$

It is easy to see that the expected value of this prediction is equal to the expected value of the frequency. Its variance consists of two components

$$\mathrm{var}(\bar{P}) = \mathrm{var}_1(\bar{P}) + \mathrm{var}_2(\bar{P}).$$

The first is due to the binomial variation of the outcomes,

$$\mathrm{var}_1(\bar{P}) = \frac{1}{N^2} \sum_i \hat{P}[\mathbf{x}_i](1 - \hat{P}(\mathbf{x}_i)),$$

and this varies with the size of the prediction sample N. The second term is the variance that is due to the use of estimated values; it is given by

$$\mathrm{var}_2(\bar{P}) = \frac{1}{N^2} \mathrm{var}\left(\sum_i \hat{P}(\mathbf{x_i})\right).$$

The variance of the sum of estimated probabilities must be derived from the variance of their common parameter estimates; we shall give a general formula for this sort of thing in Section 3.1, but its application in the present case is quite intractable. Note that this variance will vary with the size of the sample used for estimation, which is different from N.

This method is *prediction by enumeration* for a sample of N observations with known covariate values; policy effects can be simulated by changing these values and establishing the effect on $\bar{P}$. The prediction sample may have been constructed so as to represent a larger group; it may also be a random sample, and it may even be the same sample that has served for estimation. When N is large, the distribution of the covariates is more easily described by their joint density function $g(\mathbf{x}, \zeta)$ than by listing huge numbers of $\mathbf{x}_i$; the prediction is then

$$\bar{P} = \int_{\mathbf{x}} P(\mathbf{x}, \beta) \, g(\mathbf{x}, \zeta) \, \mathrm{d}\mathbf{x}.$$

It might be worth looking for a pair of functions P and g that permit explicit integration so that the prediction is at once given as a function of the parameters ζ, which are characteristics of the population under consideration like the mean and variance of the income distribution. The effect of changes in these parameters on $\bar{P}$ can then be assessed immediately. There are however not many cases where g and P mesh to produce a neat analytical expression; for exceptions, see Aitchison and Brown (1957, p. 11,39), Lancaster (1979) or McFadden and Reid (1975). But with present computing facilities enumeration is quite feasible even for large samples.

There is still one case where aggregate predictions must be based on admissible predictions of individual outcomes, namely in the *microsimulation* of interdependent dynamic processes. The standard example is the demographic development of a human population over a longer period, involving birth, ageing, marriage, parenthood and death of single individuals. These processes cannot be accurately represented by average birth rates, marriage rates and so on, and individual life histories must be simulated instead. In the same way, the working of the economic system can be represented by making representative sets of households, individuals, and firms act out their interdependent behaviour under various conditions. The seminal work in this area is Orcutt et al. (1961); for a later survey see Bergmann et al. (1980). If the probability model is embedded in such a wider model, aggregate prediction is not good enough. A given car ownership rate, for instance, must be specified by indicating which individual households have cars, for in the next round of the simulation these cars generate trips and mileage, affect other expenditure, in short set off an entire train of consequences, and this can only be followed up if car ownership is attributed to specific households. This problem is resolved by generating the random outcome by simulation according to its estimated probability.

It is often desirable to give the prediction in the form of a confidence interval or at least to have an indication of its standard error, and this would mean assessing the two sources of the variance of the overall outcome distinguished earlier in an even more complicated form. This may quickly become quite intractable, and in microsimulation the best one can do is probably to assess the intrinsic variation of the process from the variation among numerous replications of the entire simulation.

3
Maximum likelihood estimation of the binary logit model

This chapter deals with the elements of maximum likelihood estimation and with their application to the binary logit model, as practised in the logit routines of program packages. It explains the technique so that the reader can understand what goes on, and it pays particular attention to the properties of the resulting estimates that will be used in later chapters. A detailed example follows.

3.1 Principles of maximum likelihood estimation

Probability models are often estimated from survey data, which provide samples of several hundreds or even thousands of independent observations with a wide range of variation of the regressor variables. Since the advent of modern computing the preferred technique of estimation is the method of maximum likelihood. This permits the estimation of the parameters of almost any specification of the probability function. It yields estimates that are consistent and asymptotically efficient, together with estimates of their asymptotic covariance matrix and hence of the (asymptotic) standard errors of the estimates. Many statistical program packages provide ready routines for the maximum likelihood estimation of logit and probit models (see Section 1.3), but even when these are used it can be helpful to understand the first principles of the method. The present section shows how it works for binary probability models, without going into the underlying theory.

First consider the general case of any probability model. The data consist of $i = 1, 2, \ldots, n$ observations on (a) the *outcome* or occurrence of a certain event or state, represented by the $(0, 1)$ variable Y_i, and (b) a number of covariates $X_{0i}, X_{1i}, X_{2i}, \ldots$ (with the X_{0i} unit constants), which are arranged in the vector $\mathbf{x}_i$. The probability that observation i

is a *success* and has $Y_i = 1$ is

$$P_i = P(\mathbf{x}_i, \boldsymbol{\theta}), \tag{3.1}$$

with any given specification of the probability function $P(\cdot)$. It is always assumed that successive observations are independent, so that the probability density of any given ordering of observed outcomes, say $1, 0, 1, 1, \ldots$, is the product

$$P_1 \cdot Q_2 \cdot P_3 \cdot P_4 \ldots$$

The sample density of a vector $\mathbf{y}$ of zeros and ones is therefore, written in full,

$$f(\mathbf{y}, \mathbf{X}, \boldsymbol{\theta}) = P(\mathbf{x}_1, \boldsymbol{\theta}) \cdot Q(\mathbf{x}_2, \boldsymbol{\theta}) \cdot P(\mathbf{x}_3, \boldsymbol{\theta}) \cdot P(\mathbf{x}_4, \boldsymbol{\theta}) \ldots,$$

where $\mathbf{X}$ denotes a matrix arrangement of the n vectors of regressor variables $\mathbf{x}_i$ and $Q(\cdot)$ is the complement of $P(\cdot)$. In this density, the sequence of outcomes $\mathbf{y}$ is the argument, $\boldsymbol{\theta}$ is a vector of unknown fixed parameters, and the elements of $\mathbf{X}$ are known constants. The *likelihood function* L of the sample has exactly the same form, but now the sequence of zeros and ones is fixed, as given by the sample observations, and $\boldsymbol{\theta}$ is the argument. The character of $\mathbf{X}$ does not change. By strict standards a combinatorial term should have been included in the density, and hence in the likelihood, to allow for the number of permutations of the ordered observations. But this term does not contain $\boldsymbol{\theta}$ and merely adds a multiplicative constant to the density and the likelihood, or an additive constant to $\log L$; it is therefore of no consequence for the maximization of that function with respect to $\boldsymbol{\theta}$ that we shall shortly undertake, and it is ignored throughout.

We make use of the probability $\Pr(Y_i)$ of the observed outcome of Section 1.4

$$\Pr(Y_i) = P_i^{Y_i} Q_i^{1-Y_i},$$

to write the likelihood as

$$L = \prod_i \Pr(Y_i).$$

Since the probabilities $\Pr(Y_i)$ all lie between 0 and 1, so does their product L, and its logarithm will never exceed 0. This is the loglikelihood function

$$\log L(\boldsymbol{\theta}) = \sum_i \log \Pr(Y_i), \tag{3.2}$$

or

$$\log L(\boldsymbol{\theta}) = \sum_{i=1}^{n} [Y_i \log P(\mathbf{x}_i, \boldsymbol{\theta}) + (1 - Y_i) \log Q(\mathbf{x}_i, \boldsymbol{\theta})]. \qquad (3.3)$$

Another way to write $\log L$ is

$$\log L(\boldsymbol{\theta}) = \sum_{i \in \mathcal{A}_1} \log P(\mathbf{x}_i, \theta) + \sum_{i \in \mathcal{A}_0} \log Q(\mathbf{x}_i, \boldsymbol{\theta}), \qquad (3.4)$$

where $\mathcal{A}_1$ and $\mathcal{A}_0$ denote the sets of observations with Y equal to 1 and 0 respectively. The actual ordering of the observations in these expressions is immaterial; since the observations are independent, their order is arbitrary, and it does not affect their density nor the (log)likelihood. The three expressions are identical; (3.4) suggests an attractive layout of the calculations (but they are never performed by hand), while (3.3) is more convenient in the derivations which follow.

The Maximum Likelihood Estimate or MLE of $\boldsymbol{\theta}$ is $\hat{\boldsymbol{\theta}}$ which maximizes the likelihood or its logarithm; it is found by equating the derivatives of $\log L$ to zero. By convention these derivatives form a row vector; transposition yields the *score vector*, as in

$$(\partial \log L / \partial \boldsymbol{\theta})^T = \mathbf{q},$$

with typical element

$$q_j = \partial \log L(\boldsymbol{\theta}) / \partial \theta_j.$$

The estimates $\hat{\boldsymbol{\theta}}$ are obtained by solving the system of equations

$$\mathbf{q}(\hat{\boldsymbol{\theta}}) = \mathbf{0}. \qquad (3.5)$$

As a rule these equations have no analytical solution, and $\hat{\boldsymbol{\theta}}$ is found by successive approximation. One way to do this is to expand $\mathbf{q}(\boldsymbol{\theta})$ around some given $\boldsymbol{\theta}^\circ$ in the neighbourhood of $\hat{\boldsymbol{\theta}}$ in a Taylor series. This yields

$$\mathbf{q}(\hat{\boldsymbol{\theta}}) \approx \mathbf{q}(\boldsymbol{\theta}^\circ) + \mathbf{Q}(\boldsymbol{\theta}^\circ)(\hat{\boldsymbol{\theta}} - \boldsymbol{\theta}^\circ),$$

where $\mathbf{Q}$ denotes the matrix of second derivatives or Hessian matrix of $\log L$. From this we find

$$\hat{\boldsymbol{\theta}} \approx \boldsymbol{\theta}^\circ - \mathbf{Q}(\boldsymbol{\theta}^\circ)^{-1} \mathbf{q}(\boldsymbol{\theta}^\circ).$$

Since this holds only approximately, it cannot be used to determine $\hat{\boldsymbol{\theta}}$ from $\boldsymbol{\theta}^\circ$, but it can be used to obtain a closer approximation. In an iterative scheme the next approximation $\boldsymbol{\theta}_{t+1}$ is calculated from $\boldsymbol{\theta}_t$ by

$$\boldsymbol{\theta}_{t+1} = \boldsymbol{\theta}_t - \mathbf{Q}(\boldsymbol{\theta}_t)^{-1} \mathbf{q}(\boldsymbol{\theta}_t). \qquad (3.6)$$

This is known as *Newton's method*, or as the *Newton–Raphson method* or as *quadratic hill-climbing*.

The Hessian **Q** has other uses as well. Its expected value with reverse sign is the Fisher *information matrix* **H**,

$$\mathbf{H} = -E\mathbf{Q}, \tag{3.7}$$

where E takes the mathematical expectation of each element of **Q**. The inverse of **H** is the asymptotic covariance matrix of the MLE $\hat{\boldsymbol{\theta}}$

$$\mathbf{V}(\hat{\boldsymbol{\theta}}) = \mathbf{H}^{-1}. \tag{3.8}$$

The elements of **Q**, **H** and **V** are in general functions of $\boldsymbol{\theta}$; we estimate them by substituting $\hat{\boldsymbol{\theta}}$. The estimated covariance matrix is thus

$$\hat{\mathbf{V}} = \mathbf{H}(\hat{\boldsymbol{\theta}})^{-1}.$$

Recall that **H** is constructed according to (3.7). (Asymptotic) standard deviations of the parameter estimates follow immediately by taking square roots of the diagonal elements. The variance (and hence the standard deviation) of any transformation of the estimated coefficients, like the derivative (2.3) or the quasi-elasticity (2.7) of Section 2.1, or a prediction from Section 2.4, can also be obtained. For any reasonably well-behaved function $\varphi(\hat{\boldsymbol{\theta}})$ of the estimates $\hat{\boldsymbol{\theta}}$ with covariance matrix **V** we have

$$\mathrm{var}\varphi \approx \boldsymbol{\varphi}'\mathbf{V}\boldsymbol{\varphi}'^{\mathrm{T}}, \tag{3.9}$$

where $\boldsymbol{\varphi}'$ denotes the row vector of derivatives of ϕ with respect to $\boldsymbol{\theta}$. The estimated variance is of course obtained by evaluating the derivatives at $\hat{\boldsymbol{\theta}}$ and inserting $\hat{\mathbf{V}}$ for **V**.

Since we need **H** in the end, we may as well use it at an earlier stage, and substitute it for $-\mathbf{Q}$ into (3.6) which is after all only a means of generating successive approximations. This leads to the iterative scheme known as *scoring*, that is

$$\boldsymbol{\theta}_{t+1} = \boldsymbol{\theta}_t + \mathbf{H}(\boldsymbol{\theta}_t)^{-1}\mathbf{q}(\boldsymbol{\theta}_t). \tag{3.10}$$

The method has been attributed to Gauss and to Fisher. In practical computer implementations **H** is in its turn approximated by an expression based on the score vectors, so that the algorithm only requires subroutines for $\log L$ and for the score vector **q**. There are a number of other iterative techniques with equally modest requirements that also lead to ever closer approximations of $\hat{\boldsymbol{\theta}}$; the MAXLIK subroutine from the GAUSS program, for example, offers a choice of six different algorithms.

Their efficiency depends on the characteristics of the data set and the nature of the model. We shall not go into these technical details; the logit routine from a program package usually leaves the user little choice.

For the logit model the scoring method leads to particularly simple formulae, as will be shown in the next section. Still, note that the elements of $\mathbf{Q}$, $\mathbf{H}$ and $\mathbf{q}$ all consist of sums of n terms. These terms are functions of $\mathbf{x}_i$ and of $\boldsymbol{\theta}$, and as the $\mathbf{x}_i$ are usually all different, all n terms have to be calculated anew for each successive $\boldsymbol{\theta}_t$. The computations are therefore extensive, and with very large data sets – over a hundred thousand observations or so – they may require some adjustment of memory space in the program and perhaps of the computer memory itself.

All iterative schemes must be completed by *starting values* $\boldsymbol{\theta}_0$, and by a *convergence criterion* to stop the process. The judicious choice of starting values can contribute to speedy convergence, and if the analyst has a choice in the matter this is an occasion to think in advance about plausible parameter values; this will later be of great help in interpreting the results. As for the convergence criterion, the iterative process may be stopped (1) when $\log L$ ceases to increase perceptibly, (2) when the score vector approaches zero, or (3) when successive parameter values are nearly equal. Most program packages employ default convergence criteria that are absurdly strict in view of the precision of the data and of the statistical precision of the final point estimates; but this merely adds a few more iterations at negligible computing cost.

In the end, the iterative scheme yields the following results:

- MLE $\hat{\boldsymbol{\theta}}$ of the parameter vector. Under quite general conditions these estimates are consistent, asymptotically efficient, and asymptotically normal.

- Corresponding estimates of functions of $\boldsymbol{\theta}$ like derivatives and quasi-elasticities, which are also MLEs since a function of an MLE is itself an MLE of that function.

- (Asymptotic) standard errors of the parameter estimates (and perhaps of estimated functions of the parameters), derived from the estimate of their (asymptotic) covariance matrix, or from (3.9).

- The maximum value of the loglikelihood function, $\log L(\hat{\boldsymbol{\theta}})$.

The value of the loglikelihood function for particular sets of parameter estimates is useful for testing simplifying assumptions, like zero coefficients, or the absence of certain variables from the model, or other restrictions on the parameter vector $\boldsymbol{\theta}$. Provided the restricted model is *nested* as a special case within the unrestricted model, the restrictions can be tested by the loglikelihood ratio or LR test. The test statistic is

$$\mathrm{LR} = 2[\log \mathrm{L}(\hat{\boldsymbol{\theta}}_\mathrm{u}) - \log \mathrm{L}(\hat{\boldsymbol{\theta}}_\mathrm{r})], \tag{3.11}$$

with u and r denoting unrestricted and restricted parameter estimates. Under the null hypothesis that the restriction holds, this statistic is asymptotically distributed as chi-squared with r degrees of freedom, equal to the number of (independent) restrictions on the parameter vector. This is not the only test of a nested hypothesis; we return to the subject in Section 4.1.

3.2 Sampling considerations

Before we implement the general principles of maximum likelihood estimation we must clear up the role of the covariates $\mathbf{x}_i$. These were last seen in the probabilities (3.1) at the very start of this chapter, and in the loglikelihoods that followed, where they were described as 'known constants'. While they were tacitly dropped from the subsequent formulae, they are of course still part of $\mathbf{q}$, $\mathbf{Q}$, and $\mathbf{H}$.

The designation of the $\mathbf{x}_i$ as 'known constants' is ambiguous. In a laboratory experiment, the values of $\mathbf{x}_i$ are set by the analyst, in accordance with the rules of experimental design or otherwise, and they can validly be regarded as known nonrandom constants. But this argument does not hold for survey data from a genuine sample from a given population. Yet the use of the same formulae can be vindicated for this case.

A proper description of a sample survey is to consider the sample observations as drawings from a joint distribution of $\mathbf{x}$ and Y, with a density $h(\cdot)$ for a single observation. In this way we treat the sample regressor variables or covariates as random variables in their own right. The joint density $h(\cdot)$ can be written as the product of the conditional density of Y and the marginal density of $\mathbf{x}$, or

$$h(Y, \mathbf{x}, \boldsymbol{\theta}) = f(Y, \mathbf{x}, \boldsymbol{\theta}) \cdot g(\mathbf{x}, \boldsymbol{\zeta}). \tag{3.12}$$

The extended parameter vector $\boldsymbol{\eta}$ comprises both $\boldsymbol{\theta}$ and $\boldsymbol{\zeta}$. Note that the conditional density $f(\cdot)$ of Y corresponds to the initial probability (3.1). The marginal density of $\mathbf{x}$ reflects both the joint distribution of these variables in the population and the operation of the sampling method. With random sampling it is identical to the population density; in a *stratified* sample the sampling scheme deliberately distorts the population distribution by selecting elements with reference to the value of $\mathbf{x}$.

This can be spelled out by writing $h(\cdot)$ itself as the product of the population distribution of characteristics and the conditional probability that a member of that population with certain characteristics is included in the sample.

If we now follow the same passage as before from the density of a single observation (3.12) to the density of the entire sample, from the density to the likelihood, and from the likelihood to the loglikelihood, the latter is found to consist of the sum of two terms. The first represents $f(\cdot)$ and is identical to the original loglikelihood of (3.2), and the second reflects $g(\cdot)$. The total loglikelihood of the sample is then

$$\log L^* = \log L(\boldsymbol{\theta}) + \sum_i \log g(\mathbf{x}_i, \boldsymbol{\zeta}). \tag{3.13}$$

In the estimation of $\boldsymbol{\theta}$ by maximizing this function with respect to $\boldsymbol{\theta}$ the second term can be ignored, and this brings us back to the same loglikelihood function that formed the basis of estimation before. That loglikelihood and the ensuing derivations therefore hold equally well for a survey as for experimental data, but for different reasons. The survey argument is generally known as 'conditioning on the covariates'.

Closer examination of (3.13) prompts two further observations. The first is that the same data used for estimating the relationship between Y and $\mathbf{x}$ may also be informative about the distribution of $\mathbf{x}$ itself. This should cause no surprise. If $\mathbf{x}$ is income, the same sample can be used to study the relation of car ownership to income as well as the income distribution. Most sample surveys are used for several analyses with different purposes.

The second point to note is that the above argument for the use of the partial loglikelihood depends critically on the fact that the marginal distribution of $\mathbf{x}$ does not depend on $\boldsymbol{\theta}$, in other words that the sample values of $\mathbf{x}$ are independent of the values taken by Y. If this is not so, and the probability that an element of the population is included in the sample is related to the value of Y, there is *endogenous sample selection* or *state-dependent sampling*, and the present argument no longer holds. If this is disregarded it may do serious damage to the maximum likelihood estimates and their properties no longer hold. The solution is to write out the joint loglikelihood of $\mathbf{x}$ and $\mathbf{y}$ in full, allowing for the effect of the outcome on the probability that an element is drawn into the sample. As a rule this is a complicated process, but by an extraordinary and unique property of the logistic function it is much simplified for logit analyses, as we shall see in Section 6.3.

In addition to laboratory experiments and genuine samples from a larger population there is a third type of data set, namely a complete set of records, like a bank's administration of its borrowers or the client database of a credit card company. These can be treated by a rather artificial analogy as if they were samples from hypothetical superpopulations.

A final point is that maximum likelihood methodology relies heavily on asymptotic arguments, which apply by strict standards only if the sample size increases without bounds. This is a theoretical construct, which can never be realized; but it is generally held that asymptotic results will hold approximately for large samples. In some cases this can be checked. In general, however, it is an open question how large the samples must be for the asymptotic results to hold. In practice, sample sizes vary over a very wide range, from a score or so in laboratory trials to several hundreds or thousands in sample surveys in the social sciences and epidemiology, as in the car ownership example (3000 households) or the analysis of hospital admission records by Silber et al. (1995) (over 70 000 records).

A related problem is that asymptotic arguments presuppose that the data set can at least in principle be extended without bounds while the sample observations retain some of their characteristics. It is uncertain whether this must be a practical proposition or whether it is sufficient that we can imagine such an exercise. For laboratory experiments it is easily envisaged, and also for genuine samples, provided they are drawn with replacement (in practice they never are). Abstruse problems may arise if this issue is followed up in more detail, but we shall not do so here.

3.3 Estimation of the binary logit model

The principles of maximum likelihood estimation are now applied to a binary probability model, and to the logit model in particular.

First we take the loglikelihood function (3.3) of a probability model and derive expressions for $\mathbf{q}$, $\mathbf{Q}$ and $\mathbf{H}$, all in terms of the probabilities P_i and Q_i which are in turn functions of the parameters $\boldsymbol{\theta}$ (and of the constants $\mathbf{x}_i$), though these arguments are omitted. To begin with we find from (3.3) for the j th element of the score vector $\mathbf{q}$,

$$q_j = \frac{\partial \log L}{\partial \theta_j} = \sum_i \left(\frac{Y_i}{P_i} - \frac{1 - Y_i}{Q_i} \right) \frac{\partial P_i}{\partial \theta_j}. \tag{3.14}$$

For a typical element of $\mathbf{Q}$ we need the second derivative,

$$Q_{jh} = \frac{\partial^2 \log L}{\partial \theta_j \partial \theta_h}$$

$$= \sum_i \left(\frac{Y_i}{P_i} - \frac{1 - Y_i}{Q_i} \right) \frac{\partial^2 P_i}{\partial \theta_j \partial \theta_h} - \sum_i \left(\frac{Y_i}{P_i^2} + \frac{1 - Y_i}{Q_i^2} \right) \frac{\partial P_i}{\partial \theta_j} \frac{\partial P_i}{\partial \theta_h}.$$

For $\mathbf{H}$ we must reverse the sign and take the expected value of this expression, which brings about a substantial simplification: the only random variable is Y_i, and upon substituting $EY_i = P_i$ the first term vanishes altogether and the second is much simplified. We end up with

$$H_{jh} = \sum_i \frac{1}{P_i Q_i} \frac{\partial P_i \partial P_i}{\partial \theta_j \partial \theta_h}. \tag{3.15}$$

At this stage we introduce the logistic specification of the probability function, or, in its general form,

$$P_i = Pl(Z) = \exp Z / (1 + \exp Z)$$
$$= Pl(\mathbf{x}^T \boldsymbol{\beta}) = \exp \left(\mathbf{x}^T \boldsymbol{\beta} \right) / \left[1 + \exp \left(\mathbf{x}^T \boldsymbol{\beta} \right) \right].$$

We have

$$dPl_i / dZ_i = Pl_i Ql_i,$$

and

$$(\partial Z_i / \partial \boldsymbol{\beta})^T = \mathbf{x}_i.$$

As before, transposition is in order since the derivatives of a scalar with respect to a vector argument are conventionally arranged in a row vector. – Upon substitution of these expressions by the chain rule into (3.14) we obtain the score vector

$$\mathbf{q} = \sum_i \left(\frac{Y_i}{P_i} - \frac{1 - Y_i}{Q_i} \right) Pl_i Ql_i \mathbf{x}_i$$

$$= \sum_i \left(Y_i Ql_i - (1 - Y_i) Pl_i \right) \mathbf{x}_i$$

$$= \sum_i \left(Y_i - Pl_i \right) \mathbf{x}_i. \tag{3.16}$$

Substitution of the same expressions into (3.7) yields an equally simple expression for the information matrix, namely

$$\mathbf{H} = \sum_i Pl_i Ql_i \mathbf{x}_i \mathbf{x}_i^T. \tag{3.17}$$

We recall that Pl_i and Ql_i are functions of the values assigned to the parameter vector $\boldsymbol{\theta}$ (as well as of the $\mathbf{x}_i$). Upon restoring these arguments and substituting the result in (3.10), the iterative scheme is written in full as

$$\boldsymbol{\theta}_{t+1} = \boldsymbol{\theta}_t + \left[\sum_i Pl_i\left(\boldsymbol{\theta}_t\right) Ql_i\left(\boldsymbol{\theta}_t\right) \mathbf{x}_i\mathbf{x}_i^T\right]^{-1} \sum_i [Y_i - Pl_i\boldsymbol{\theta}_t)]\mathbf{x}_i. \quad (3.18)$$

This is the scoring method of determining the MLE of the parameters of the logit model by successive approximation.

Bearing in mind that the regressor vectors $\mathbf{x}_i$ form the rows of $\mathbf{X}$, so that

$$\sum_i \mathbf{x}_i\mathbf{x}_i^T = \mathbf{X}^T\mathbf{X},$$

the Hessian (3.17) bears a close resemblance to the Hessian of the linear regression model,

$$(\mathbf{X}^T\mathbf{X})\sigma^2.$$

The difference lies only in the weights Pl_iQl_i, which represent the variance of Y_i, just as σ^2 does in ordinary linear regression. Several considerations from the linear regression model carry over.† To begin with, $\mathbf{X}$ must have full column rank, and strong but imperfect collinearity of the regressors will lead to a near-singular information matrix and to large elements of the covariance matrix (3.8), and hence to large standard errors of the estimates. In the present case we have an additional interest in a well-conditioned matrix $\mathbf{H}$ since any numerical difficulties that may arise in its inversion will affect the speed and efficiency of the iterative process. Apart from not being near-singular, $\mathbf{X}^T\mathbf{X}$ should preferably be well balanced in the sense that its elements do not vary too much in absolute size. This can be ensured by appropriate scaling of the regressor variables.

Once the MLE $\hat{\boldsymbol{\theta}}$ has been obtained we may consider the estimated or predicted sample probabilities

$$\widehat{Pl}_i = Pl(\mathbf{x}_i, \hat{\boldsymbol{\theta}}). \quad (3.19)$$

At the MLE $\hat{\boldsymbol{\theta}}$ the score vector must satisfy (3.5), so that by (3.16)

$$\sum_i (Y_i - \widehat{Pl}_i)\mathbf{x}_i = \mathbf{0}. \quad (3.20)$$

---

† Note that the solution of (3.18) cannot be represented as a weighted linear regression problem, since the weights $Pl_i(\boldsymbol{\theta})Ql_i(\boldsymbol{\theta})$ depend on the parameter values and must be adjusted in every round of the iterative process.

The first term on the left can be defined as a *(logit) quasi-residual*

$$e_{li} = Y_i - \widehat{Pl}_i \tag{3.21}$$

and $\mathbf{e}_l$ as the vector of quasi-residuals, so that in self-evident notation

$$\mathbf{y} = \hat{\mathbf{p}} + \mathbf{e}_l. \tag{3.22}$$

From (3.21) we have

$$\mathbf{X}^T \mathbf{e}_l = \mathbf{0}. \tag{3.23}$$

This is a direct analogue of

$$\mathbf{X}^T \mathbf{e} = \mathbf{0}$$

for the ordinary residuals of OLS regression, which in that model leads to the *normal equations*

$$\mathbf{X}^T \mathbf{X} \mathbf{b} = \mathbf{X}^T \mathbf{y}.$$

The quasi-residuals e_{li} do indeed share several properties of the OLS residuals e_i. Like these, they represent the difference between the observed outcome and the estimate of its expected value, and have expectation zero; but the logit quasi-residuals are constrained to the interval $(-1, +1)$, and they are intrinsically heteroskedastic as their variance is

$$Pl_i Ql_i.$$

This suggests the use of standardized (quasi-)residuals with variance 1

$$\tilde{e}_{li} = e_{li} / \sqrt{\hat{P}_i \hat{Q}_i} \tag{3.24}$$

also known as *Pearson* or *Studentized* residuals, since they are scaled by estimated probabilities. – Note that the quasi-residuals do not in any sense correspond to the random disturbances ε_i of the underlying regression equation (2.8) of Section 2.3.

If β contains an intercept (as it invariably does), the $\mathbf{x}_i$ contain a unit element, and it follows from (3.23) that

$$\imath^T \mathbf{e}_l = \sum e_{li} = 0,$$

where $\imath$ is the unit vector. Thus the quasi-residuals sum to zero, again just like OLS residuals. This can be rewritten as

$$\sum_i (Y_i - \widehat{Pl}_i)/n = 0,$$

or

$$\sum_i \widehat{Pl}_i/n = \sum_i Y_i/n. \tag{3.25}$$

This is the *equality of the means*: the mean predicted probability of the sample equals the actual sample relative frequency. It comes about because the score vector of the logit model is linear in $\mathbf{x}_i$.

This property also holds for a logit model with an intercept only, the *null model* or *base-line model*. This logistic regression with all slope coefficients constrained to zero can be used as a benchmark in an LR test of the significance of the set of covariates. In the base-line model, the argument $\mathbf{x}_i^T \boldsymbol{\beta}$ is reduced to a single constant β_0, and the same probabilities apply to all observations, namely

$$Pl^\circ = \exp \beta_0 / (1 + \exp \beta_0),$$

$$Ql^\circ = 1 - Pl^\circ = 1/(1 + \exp \beta_0).$$

By (3.25), the maximum likelihood estimate $\hat{\beta}_0$ must satisfy

$$\widehat{Pl}^\circ = \sum_i Y_i/n = m/n = f,$$

with m the number of sample observations with $Y_i = 1$ or success, and f its relative frequency in the sample. It follows that $\hat{\beta}_0$ is the logit or log-odds of this relative sample frequency,

$$\hat{\beta}_0 = R(\widehat{Pl}^\circ) = \log \widehat{Pl}^\circ / \widehat{Ql}^\circ = \log [f/(1-f)]$$

$$= \log m - \log(n - m). \tag{3.26}$$

The corresponding null loglikelihood is

$$\log L^\circ = m \log f + (n - m) \log(1 - f)$$

$$= m \log m + (n - m) \log(n - m) - n \log n. \tag{3.27}$$

The null model is a restricted version of the model with argument $\mathbf{x}^T \boldsymbol{\beta}$, all slope coefficients being equated to zero. The corresponding LR test statistic (3.11) is

$$\mathrm{LR}^\circ = 2[\log L(\hat{\boldsymbol{\beta}}) - \log L^\circ], \tag{3.28}$$

and under the null hypothesis that the regressor variables have no effect this has a chi-squared distribution with k degrees of freedom, with k the number of slope coefficients in the unrestricted model.

We return to the practice of estimation. The iterative scheme (3.18) must still be supplemented by *starting values* $\boldsymbol{\beta}_0$ of the parameters, and by a *convergence criterion* to stop the process. It will be shown in Section 7.2 that for the logit model the scoring algorithm always converges to a single maximum in the end, so that any starting value will do: its choice is important only for the speed of convergence, not for the final result. Awkward starting values may however still occasionally cause computational difficulties. Program routines may set all parameters equal to zero, which implies that in the first round all P_i are set equal to 0.5; another method is to put all slope coefficients equal to zero and the intercept at its estimate of (3.26), equating all P_i to the sample frequency.

With reasonable data there should be no problems in running a logit routine from a program package. In the case of perfect multicollinearity the program usually protests by sending an error message indicating overflow or division by zero or the like, and the same holds in case the *zero cell* complication described in the next section occurs. Convergence should be achieved in something like five or, at the outside, ten iterations; if the number is much larger, something is wrong. First, the data may be ill conditioned, with an almost singular regressor matrix, or with covariates of widely different orders of magnitude (as measured by their variance). If the regressors are severely collinear, one or two can be omitted; if they show too little variation, other regressors must be sought; if the regressor matrix is unbalanced, rescaling the variables will help. The convergence criterion is seldom at fault; it is usually much too strict by any reasonable standards, but this will at worst cause a number of unnecessary iterations. If necessary the convergence criterion built into a program package can sometimes be circumvented by making the program print output at each iteration, and stopping the process by hand when the loglikelihood or the parameter values cease to vary in successive iterations.

3.4 Consequences of a binary covariate

Two special cases worth mentioning arise if there is a regressor that is itself a discrete binary variable which only takes the values 0 or 1. The first is an unpleasant complication of the estimation process and the second opens the way to case–control studies.

For the estimation problem we recall that for any particular regressor X_j from among many (3.23) implies

$$\sum_i X_{ji} e_{li} = 0,$$

or, partitioning the observations according to the outcome, as in (3.4),

$$\sum_{i \in \mathcal{A}_1} X_{ji} e_{li} + \sum_{i \in \mathcal{A}_0} X_{ji} e_{li} = 0.$$

Since $e_{li} = Y_i - \hat{P}_i$ and $0 < \hat{P}_i < 1$, all e_{li} of the first term are positive and those of the second term are negative. In the special case that X_j is itself a binary discrete variable taking the values 0 and 1 only, the first sum is nonnegative and the second sum nonpositive. But in the *very* special case that X_{ji} happens to be 0 for *all* observations with $Y_i = 0$, the second term is 0. The first term will then be positive, for all its e_{li} are positive and at least some of its X_{ji} must be 1 (obviously not all X_{ji} can be 0). In other words, this element of the score vector cannot be 0, and (3.5) has no solution. In this very special case the derivative of $\log L$ with respect to the slope coefficient β_j is therefore always positive, the algorithm that seeks to maximize $\log L$ attaches ever increasing values to this coefficient, and the maximum likelihood method breaks down.

In the example that we have presented X_{ji} is 0 for all observations with $Y_i = 0$, and in a 2×2 table of frequencies by X_j and Y the cell for $(X_j = 1, Y = 0)$ is empty. This complication is therefore known as the *zero cell* defect of the data. It may also arise in a less obvious manner, for example with an X_j that is either 0 or positive, not necessarily 1, that is again 0 for all observations with the same outcome. The presence of one covariate with these particular properties among many is enough to play havoc with the estimation routines. If there is a cell with very small but nonzero frequency, estimation will not break down, but the quality of the estimates will be unfavourably affected. If this defect occurs, inspection must show whether it reflects a systematic effect, and hence is of substantive interest, or whether it is merely an accident of sampling or of recording the observations. In most cases the offending covariate will have to be deleted. Agresti(1966, p. 190 ff.) and Menard(1995, p. 67 ff.) give more details.

The second special case arises if a binary discrete variable X is the *only* regressor variable in addition to the unit constant. The argument of the logit model can be written as

$$\alpha + \beta X_i,$$

and the $\mathbf{x}_i$ consist of two elements and come in two varieties only, namely $[\,1\ 1\,]^T$ and $[\,1\ 0\,]^T$. The Pl_i take only two values, namely $Pl(\alpha + \beta)$ in the first case and $Pl(\alpha)$ in the second. As a result, the score vector $\mathbf{q}$ of (3.16) consists of two elements that are easily expressed in terms of these two probabilities and of the four frequencies of observations for the double dichotomy $Y_i = 0, 1$, $X_i = 0, 1$. Upon equating q to zero we have two equations in two unknowns (the two probabilities) that can be solved directly; and from these probabilities follow the parameters α and β. This is one of the rare exceptions that the maximum condition $q = 0$ does have a simple analytical solution. This forms the basis of the estimation of the odds ratio or the log odds ratio (which is β) in the case–control studies of Section 6.4.

3.5 Estimation from categorical data

Survey data often contain variables that are categorical, either by their nature, like gender or nationality, or by convention as in variables like 'level of education' or 'degree of urbanization', which are quite capable of continuous variation but which are limited to a few broad classes by the way they have been recorded or by a deliberate choice of the analyst. Sensitive variables like income are sometimes determined by asking respondents to indicate their position on a list of fairly broad intervals; for items like habitat and education it makes sense to adopt the standard classification. In the past survey observations were also often grouped into classes and allotted group means in order to simplify computations. In other fields like epidemiology many variables are categorical by nature, too, like 'treatment' or 'diagnosis'. If *all* relevant covariates are such discrete categorical variables, we may set up a complete cross tabulation of the sample with a limited number of cells; the total sample information is then summarized by the number of observations with and without the outcome attribute for each cell. Such categorical or *grouped* data permit a greater choice of methods of estimation.

Let $j = 1, 2, \ldots, J$ denote the cells defined by the cross-classification of the sample by the categorical covariates. Empty cells, which are bound to occur if we cross all available classifications, are ignored. The number of observations in each cell is n_j, the number of successes with $Y_i = 1$ is m_j, and the number with $Y_i = 0$ is $l_j = n_j - m_j$; the relative frequency of the attribute in cell j is

$$f_j = m_j / n_j.$$

The covariates take a single value for each cell; this may be a $(0, 1)$ dummy, a class rank on a conventional scale, or the mid-class value or class mean of a continuous variable, as the case may be. All within-cell variation of these variables is ignored.

A straightforward method of estimation is to regard the grouped observations simply as repeated individual observations, and to apply the methods of the preceding section. This is merely a matter of adjusting the notation. From (3.16) we find for the score vector

$$\mathbf{q} = \sum_j [m_j - n_j Pl_j(\boldsymbol{\theta})]\mathbf{x}_j$$

$$= \sum_j n_j [f_j - Pl_j(\boldsymbol{\theta})]\mathbf{x}_j.$$

The equivalent of the maximum condition (3.19) is therefore

$$\sum_j n_j (f_j - \widehat{Pl}_j)\mathbf{x}_j = \mathbf{0},$$

with $\widehat{Pl}_j = Pl(\mathbf{x}_j, \hat{\boldsymbol{\theta}})$. The discrepancy $(f_j - \widehat{Pl}_j)$ will asymptotically tend to zero: as all n_j increase, along with the sample size, it is basic sampling theory that f_j converges to Pl_j, while $\widehat{Pl}_j$ converges to Pl_j too since $\hat{\boldsymbol{\theta}}$ is a consistent estimate of $\boldsymbol{\theta}$, and therefore $\widehat{Pl}_j$ is a consistent estimate of Pl_j.

For $\mathbf{H}$ of (3.17) we find

$$\mathbf{H} = \sum_j n_j Pl_j Ql_j \mathbf{x}_j \mathbf{x}_j^T.$$

Together, these expressions permit the implementation of the scoring algorithm (3.10), viz.

$$\boldsymbol{\theta}_{t+1} = \boldsymbol{\theta}_t + \mathbf{H}(\boldsymbol{\theta}_t)^{-1}\mathbf{q}(\boldsymbol{\theta}_t).$$

A different method of estimation is based on the chi-squared statistic for testing whether an observed frequency distribution agrees with a given probability distribution. The general formula of this classic Pearson goodness-of-fit statistic is

$$\chi^2 = \sum_j \frac{(s_j - \hat{s}_j)^2}{\hat{s}_j},$$

where s_j and $\hat{s}_j$ denote the observed and the predicted frequencies in class j. This nonnegative quantity has a chi-squared distribution with $J - K$ degrees of freedom, with J the number of cells and K the number of adjusted parameters that enter into the predicted numbers, provided the numbers in the cells are not too small: the standard prescription is that the predicted frequencies must not be smaller than 5. For large values of this statistic we must reject the null hypothesis that the s_j are a sample from the given distribution.

In the present case the predicted frequencies are those given by a logit model with estimated or assumed parameter values. Consider a single cell of the cross-classification. With two classes only (with and without the attribute) and frequencies m_j and l_j, $m_j + l_j = n_j$, the test statistic is

$$\chi_j^2 = \frac{(m_j - \hat{m}_j)^2}{\hat{m}_j} + \frac{(l_j - \hat{l}_j)^2}{\hat{l}_j},$$

or, after some rearrangement

$$\chi_j^2 = n_j \frac{(f_j - \hat{f}_j)^2}{\hat{f}_j(1 - \hat{f}_j)}.$$

These terms sum over all cells to

$$\chi^2 = \sum_j n_j \frac{(f_j - \hat{f}_j)^2}{\hat{f}_j(1 - \hat{f}_j)}. \tag{3.29}$$

The null hypothesis is that the observed cell dichotomies have been generated by the logit model; for large values of the last statistic this must be rejected. Conversely this statistic can be taken as a criterion function that must be minimized for purposes of estimation, thus defining *minimum chi-squared estimates* of $\boldsymbol{\theta}$ (which enters into the formula via the predicted $\hat{m}_j$ and $\hat{l}_j$). Here, too, zero or very small values of the cell numbers m_j and l_j, $\hat{m}_j$ and $\hat{l}_j$, should be avoided, if necessary by pooling adjacent cells of the original classification by covariates. This estimator is not identical to the maximum likelihood estimator, although the two share the same asymptotic properties; see Rao (1955). Minimum chi-squared estimation has been advocated with fervour by Berkson (1980), who stresses its superior small-sample qualities.

There is a third method of estimation from categorical data; once more zero (or very small) values of m_j or l_j must be avoided. In that case the relative frequencies f_j do not attain the bounds of 0 or 1, and we may apply the log-odds transformation of (2.6)

$$\text{logit}(f_j) = \log[f_j/(1 - f_j)].$$

This is the original meaning of the logit transformation. Insofar as f_j is approximately equal to the logistic probability $Pl(\mathbf{x}_j^T \boldsymbol{\beta})$, we have

$$\text{logit}(f_j) \approx \mathbf{x}_j^T \boldsymbol{\beta}.$$

This immediately suggests plotting logit(f_j) against X or fitting an OLS regression. A probit transformation of the frequencies, obtained from tabulations, can be used in the same manner. More about these techniques from the old days, before the advent of the computer, can be found in Chapter 9.

3.6 Private car ownership and household income

We illustrate the binary logit model by an analysis of private car ownership in a Dutch household survey of 1980. The data have been described in Section 1.3. In this chapter we consider a single attribute, namely ownership of one or more private cars, new or used, and in this section the log of household income per head or LINC is the only explanatory variable. We thus revert to the simple model

$$\Pr(Y_i = 1) = Pl(\alpha + \beta \text{ LINC}_i).$$

For economists this is an example of an Engel curve, although of a peculiar variety as income determines ownership and not expenditure on a particular good.

To begin with α and β have been estimated by the scoring method of (3.18), as embodied in an early version of the LOGITJD module of PCGIVE. Three iterations are sufficient for convergence; their course is shown Table 3.1. The first or zero iteration shows the starting values with α_0 given by the base-line estimate of (3.24) and β_0 equal to zero: all probabilities are equal to the average sample frequency of 0.65. In the next iterations the coefficients are adjusted, the loglikelihood increases, and the elements of the scoring vector $\mathbf{q}$ move towards zero. Convergence

Table 3.1. *Household car ownership and income:*
iterative estimation.

iteration	$\log L$	$\hat{\alpha}$	$\hat{\beta}$	$|q_\alpha|$	$|q_\beta|$
0	-1839.63	0.5834	0	$< 10^{-11}$	48.25
1	-1831.29	-2.7420	0.3441	2.30	22.72
2	-1831.29	-2.7723	0.3476	0.0016	0.0166
3	-1831.29	-2.7723	0.3476	$< 10^{-8}$	$< 10^{-7}$
		(3.35)	(4.03)		

Absolute value of t − ratios in brackets.

Table 3.2. *Household car ownership by income classes.*

class limits	x_j^a	n_j	m_j	f_j	$\mathrm{logit}(f_j)$
$< 10\,000$	$7\,000$	400	220	0.55	0.20
$10\,000 - 15\,000$	$13\,000$	962	627	0.65	0.63
$15\,000 - 25\,000$	$20\,000$	992	636	0.64	0.58
$25\,000 - 35\,000$	$28\,000$	330	227	0.69	0.79
$\geq 35\,000$	$40\,000$	136	100	0.74	1.02

a *Class means of* INC, *fl. p.a.*

is reached in only three iterations, but then the loglikelihood at its maximum is not very much larger than at the zero iteration. In the last line we have made use of the asymptotic standard errors, obtained from the asymptotic covariance matrix, to calculate t-values; these are shown in absolute value in brackets below the estimates to which they refer. We adhere to the common convention of treating coefficients with a t-value over 2 as significantly different from zero, or statistically significant for short.

The same relation can be estimated from grouped data. The households have been classified into five classes by INC, and the entire sample information is thus condensed in Table 3.2.† With some adjustments, mid-class values have been assigned to the classes and their logarithms are the regressor variable. Maximum likelihood estimation treating the group values as if they were repeated individual observations yields the results of Table 3.3. We find much the same result as from the individual observations, again in three iterations.

Finally, the logit transformation of the relative frequencies from Table 3.2 is plotted against the logarithm of mid-class income in Figure 3.1. Upon fitting a straight line by eye we read off its parameters as $\alpha = -3.60$, $\beta = 0.43$. The difference from the earlier estimates is that the class means have not been weighted by class numbers. The estimates of the intercept and the slope are apparently strongly correlated, for one goes down and the other up. The technique does not provide a standard error of the estimates; it only serves for a quick inspection of the data, and perhaps for finding starting values of the parameters.

The three estimation techniques yield virtually the same results, but these results are disappointing. We have already noted that $\log L$ does

† The classification by income per head is unusual; it is used here in order that the grouped data have the same regressor variable as the individual data.

Table 3.3. *Household car ownership: iterative estimation, grouped data.*

iteration	$\log L$	α	β	$|q_\alpha|$	$|q_\beta|$
0	-1839.63	0.5834	0	$< 10^{-13}$	48.38
1	-1830.89	-2.9184	0.3617	2.4144	23.33
2	-1830.88	-2.9153	0.3618	0.0013	0.0124
3	-1830.88	-2.9154	0.3618	$< 10^{-9}$	$< 10^{-8}$
		(3.48)	(4.17)		

Absolute values of t − ratios in brackets.

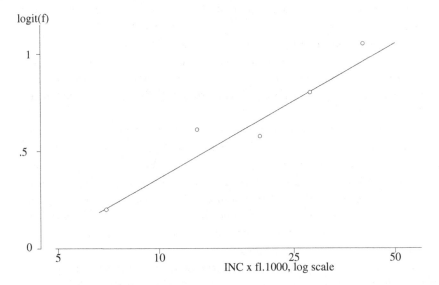

Fig. 3.1. Logit of car ownership plotted against the logarithm of income per head.

not improve much beyond the zero iteration, which is the null model of constant probability. This means that variations in income contribute very little to an explanation of private car ownership. The restrictive hypothesis of zero β may be tested by the LR test of (3.27); upon comparing the final loglikelihood of Table 3.1 with the null value, we find a test statistic of 16.68. This is quite significant, for the 5% critical value of chi-squared with one degree of freedom is 3.84. The absence of any income effect is thus rejected. We could have known, for the estimated

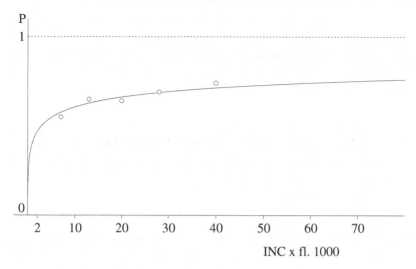

Fig. 3.2. Fitted probability of car ownership as a function of income per head.

income coefficient of Table 3.1 differs significantly from zero by its *t*-ratio of just over 4.

But this purely statistical reassurance does not alter the fact that the effect of income alone on private car ownership is small. The grouped data demonstrate directly that the incidence of car ownership varies very little from one income class to another.† Since the income variable is in logarithms, the quasi-elasticity is given by the derivative (2.3); at the sample mean ownership rate of 65%, this gives $0.35 \times 0.65 \times 0.35 = 0.08$, which is much lower than one would expect: a 12% rise in income is needed to increase the probability of car ownership by 1 percentage point. By a statistical assessment the income effect is significantly different from zero, but by economic considerations it is very weak. This is once more demonstrated by the fitted curve of private car ownership as a function of income per head in Figure 3.2. Note that the logarithmic transformation of income makes the sigmoid curve asymmetrical. The main conclusion from the graph is that its slope is very small.

There are two reasons for this low income effect. It is possible that private car ownership was already so well established in 1980 that it was indeed hardly related to income. This is unlikely, for it would mean

† Another consequence of the limited ownership variation between income classes, coupled with anomalous low car ownership in income class 3, is that the grouped data estimates are quite sensitive to the mid-class income values assigned to each class and to the treatment of the open-ended classes at either end of the range.

that the ownership levels of about 0.7 already constitute saturation. The second possibility is that the estimate of the income coefficient is biased downward by the omission of other covariates from the analysis. This is strongly suggested by the results of the next section. We return at greater length to this bias in Section 5.3.

3.7 Further analysis of private car ownership

We now add other regressors to the income variable LINC, namely all the other variables listed in Section 1.3,

- LSIZE, the log of household size;
- BUSCAR, a $(0, 1)$ dummy variable for the presence of a business car;
- URBA, the degree of urbanization, graded from countryside (1) to city (6);
- AGE, the age of the head of household by five-year classes.

We have some misgivings about measuring household size in equivalent adults, not in persons, but the definition of the variables is dictated by the available data set. Note that the order of the variables is different from their order in the data set.

The addition of significant regressor variables improves the fit, increases the precision of the estimates, and leads to higher estimates of the income effect. These effects are shown in Table 3.4. The maximum loglikelihood increases upon the addition of each new variable, as is only natural. Whether these increases are significant can be tested by the likelihood ratio test of (3.11), for at each stage the simpler model is a restricted form of the next model, with one coefficient equal to zero; since chi-squared with one degree of freedom is significant at the 5% level when it exceeds 3.84, the loglikelihoods should increase by at least half of this, or by 1.92. All additional variables pass this test, with URBA giving the weakest performance. It also stands to reason that the precision of the estimates, as reflected in their t-values, increases with each additional regressor.

Table 3.4 also shows the effect of introducing additional variables on the estimated slope coefficients and in particular on the income coefficient. This shows an increase in absolute value, away from zero, that is roughly in line with the increase in $\log L$. The effect of adding the size variable on the income coefficient is spectacular; this is due to the fact that LINC and LSIZE are negatively correlated while they have a similar effect on car ownership. If LSIZE is absent, its effect is confounded with

Table 3.4. *Household car ownership: effect of adding covariates on estimated slope coefficients.*

log L	LINC	LSIZE	BUSCAR	AGE	URBA
−1831.29	0.35				
	(4.0)				
−1614.92	1.77	2.22			
	(14.1)	(18.7)			
−1393.74	2.46	3.09	−2.95		
	(16.7)	(21.6)	(18.9)		
−1360.23	2.36	2.83	−3.00	−0.12	
	(16.1)	(19.8)	(19.4)	(8.1)	
−1351.39	2.38	2.76	−3.04	−0.13	−0.12
	(16.2)	(19.1)	(19.5)	(8.2)	(4.2)

Absolute values of t − ratios in brackets.

opposite sign with the effect of LINC, and this depresses the income coefficient. A further advance in the income coefficient takes place when the presence of a business car is taken into account. From the third line on, the improvement of log L as well as the changes in the slope are less pronounced. The mechanism of the omitted variables bias and its operation in the present case are discussed at greater length in Section 5.3.

The last line of Table 3.4 gives the best estimates that can be obtained from the present data set. Since LINC and LSIZE are in logarithms, quasi-elasticities equal derivatives, and these can be evaluated at the mean ownership level, as before; the quasi-elasticity in respect of income is then 0.54, and in respect of household size 0.63. The latter elasticity refers to a pure size effect, measured at a given income per head or a constant standard of living, and it is therefore considerably higher than the effect of an increase in family size at a given nominal income level. These are reasonable values. The presence of a business car is the third important determinant: it reduces the probability of private car ownership substantially. For otherwise identical households, the odds ratio of private car ownership for the presence of a business car is $\exp(-3) = 0.05$; this means that a business car is an almost perfect substitute for a private car. The two remaining variables are less important (though still significant): upon taking derivatives we find that moving up one step in a six-class scale of urbanization reduces the probability of private car ownership by 3 percentage points, and so does a five years older head of the household.

4

Some statistical tests and measures of fit

In discussing the car ownership example of Chapter 3 we loosely applied
tests of significance to individual coefficients and to the model as a whole.
The subject deserves a more systematic treatment and this is given in
the first section of this chapter. Textbook etiquette also demands that
estimation is accompanied by diagnostic testing, putting the model under heavy strain before it is accepted. In applied work this is not always
done. If the statistical analysis serves only for a preliminary screening of
the evidence, preparatory to substantive studies, probability models are
used in the same way as ordinary regression as an exploratory technique,
without much regard for the suitability of the underlying assumptions.
But when it comes to classification or conditional prediction with heavy
consequences, it is worth while to perform a goodness-of-fit test of the
model, and to assess its quality in describing the data.

4.1 Statistical tests in maximum likelihood theory

One of the basic requisites of the statistical model of maximum likelihood
theory is the definition of a *parameter space* in which the true parameter
vector θ as well as its MLE $\hat{\theta}$ must lie. In the context of maximum
likelihood theory, statistical tests bear on *nested* hypotheses that restrict
θ to a subspace of a wider but still acceptable parameter space. The
restriction may constrain one element of θ to a particular value (often
zero), but it may also take a more general form and involve any number
of parameters. There are three types of tests of such restrictions against
the wider alternative, namely
- LR or Likelihood Ratio tests, based on a comparison of the maximum
  value of the loglikelihood with and without restrictions;
- Wald tests, based on a comparison of the restricted parameter values

56

with the asymptotic normal distribution of the unrestricted parameter estimates;

• LM or Lagrange Multiplier tests, also known as score tests, based on a comparison of the score vector of the unrestricted model, evaluated at the constrained parameter estimates, with the unrestricted value (which is zero).

Any nested hypothesis can be subjected to all three tests. They are asymptotically equivalent, but they may give different results in any particular instance and they may have different small sample performances in specific applications.

In the car ownership analysis of Chapter 3 the relevance of covariates was assessed by testing (and rejecting) the restrictive hypothesis of a zero coefficient by LR and by Wald tests. The latter contrast the constrained value of a parameter β_j° with the distribution of its unrestricted estimate, which is a normal or rather a Student distribution. This gives

$$\frac{\hat{\beta}_j - \beta_j^\circ}{\text{s.d.}(\hat{\beta}_j)};$$

if β_j° is zero (a common choice), this is the t-value of $\hat{\beta}_j$. The natural choice for testing the joint relevance of several covariates is an LR test, and this may also serve to establish the significance of the entire model by a comparison with the null model, as in the test statistic (3.28) of Section 3.3. This last test is like the F test of ordinary regression; the null hypothesis is that the fitted model is not significantly different from the base-line model. This is almost invariably rejected, but it only means that the fitted model is better than nothing. The goodness-of-fit tests discussed below are much more stringent.

Apart from these standard tests for the direct effect of covariates there is a variety of diagnostic tests of the model against alternatives like additional transformations of the covariates (including interaction terms), heteroskedasticity, or an altogether more general model. When such issues arise the first concern is to set up a nested hypothesis, and the next step is to choose one of the three types of test. In the past this choice has been influenced strongly by considerations of computational convenience: up to the 1980s the iterative nature of maximum likelihood estimation was regarded as so onerous that Wald or score tests were preferred since they require one fit while an LR test requires two. This led authors like Pregibon (1982) and Davidson and McKinnon (1984) to develop score tests of great ingenuity and elegance; in Davidson and

McKinnon (1993) the last authors advocate their system of auxiliary regressions as a general approach to hypothesis testing. In their earlier paper, they examine the performance of several variants of these score tests in small and moderate samples (from 50 to 200 observations); the LR test, used as a standard of comparison, performs well, but is dismissed because of its computational burden. This is one of the rare studies of small sample properties of these various tests; another example is the early paper of Hauck and Donner (1977), who show that for case–control analyses of small samples the Wald test may exhibit perverse behaviour. So far LR tests have not attracted this sort of criticism. With present computing facilities they are no longer at a disadvantage, although there may still be technical difficulties in fitting both the restricted and the unrestricted model by a standard routine.

4.2 The case of categorical covariates

The case of categorical data samples provides a good illustration of the nature of goodness-of-fit tests of a model. The data consist of frequencies of the two outcomes in the homogeneous cells of a multidimensional cross-classification of the individual observations by their covariates, as discussed earlier in Section 3.5. The classification has J cells with index j, there are n_j observations in a cell, m_j successes with $Y = 1$ and l_j failures with $Y = 0$; the sample numbers are n, m and l. As before, empty cells are ignored, and zero values of m_j or l_j avoided by pooling adjacent cells. The observed relative cell frequencies of success are $f_j = m_j/n_j$, and these are contrasted with predicted or theoretical frequencies $\hat{f}_j$. In the present case these are the estimated probabilities $\hat{P}_j$ from a probability model (or more specifically a logit model) that has been fitted to the same data.

The Pearson goodness-of-fit statistic for a single cell j is

$$\chi_j^2 = \frac{(m_j - n_j \hat{f}_j)^2}{n_j \hat{f}_j}$$

and in this binary case this can be simplified to an expression in the relative frequency of success alone,

$$\chi_j^2 = \frac{(f_j - \hat{f}_j)^2}{\hat{f}_j(1 - \hat{f}_j)}.$$

The sum over all cells is then

$$\chi^2 = \sum_j n_j \frac{(f_j - \hat{f}_j)^2}{\hat{f}_j(1 - \hat{f}_j)},$$

or

$$\chi^2 = \sum_j n_j \frac{(f_j - \hat{P}_j)^2}{\hat{P}_j(1 - \hat{P}_j)}, \tag{4.1}$$

or the sum of squared Studentized residuals in each cell. This is the same expression as (3.29) of Section 3.5, where it was introduced as a criterion function for estimating the model parameters; originally, however, its use is to test the goodness-of-fit of a given model, i.e. the null hypothesis that the observed cell dichotomies have been generated by a probability distribution, here: by the fitted model. Under this hypothesis, and provided the parameter estimates are consistent (as maximum likelihood estimates are), the test statistic is chi-squared distributed with $J - k - 1$ degrees of freedom, with J the number of cells and $k + 1$ the number of adjusted parameters, viz. k slope coefficients and an intercept. For large values of (4.1) the null hypothesis must be rejected.

Another test of model performance is a likelihood ratio test in which the fitted model is regarded as a restricted variant of the *saturated* model that provides a perfect fit. In the saturated model the predicted probabilities for each cell equal the observed relative frequencies; see Bishop et al. (1975, pp. 125–126). This means putting $\hat{f}_j = f_j$; altogether J parameters are estimated, and all available degrees of freedom are used up. The saturated loglikelihood is

$$\log L^s = \sum_j [m_j \log \hat{f}_j + l_j \log(1 - \hat{f}_j)]$$

$$= \sum_j (m_j \log m_j + l_j \log l_j - n_j \log n_j),$$

and this is indeed a *maximum maximorum*. With a little ingenuity it must be possible to represent the fitted model as a restricted version that is nested within the saturated model. Its loglikelihood is

$$\log \hat{L} = \sum_j [m_j \log \hat{P}_j + l_j \log(1 - \hat{P}_j)]$$

and the LR test statistic is then

$$\mathrm{LR}^* = 2(\log L^s - \log \hat{L}). \tag{4.2}$$

This is again chi-squared distributed with $J - k - 1$ degrees of freedom. It tests whether the fitted model falls significantly short of the saturated model.

Both test statistics (4.1) and (4.2) are based on a comparison of frequencies and predicted probabilities; some rearrangement yields

$$\chi^2 = \sum_j n_j (f_j - \hat{P}_j)^2 \hat{P}_j^{-1}$$

$$\text{LR}^* = \sum_j \{ m_j \log[f_j / \hat{P}_j]^2 + l_j \log[(1 - f_j)/(1 - \hat{P}_j)]^2 \}.$$

Both quantities are asymptotically distributed as chi-squared with $J - k - 1$ degrees of freedom, regardless of sample size. Asymptotic theory is here based on increasing the sample size n for a given classification with a constant number of cells J. In this process all n_j will increase in line with n; but this effect on the test statistics is offset by the convergence of frequencies and probabilities, for the f_j tend to the true probabilities by the law of large numbers, and the $\hat{P}_j$ tend to the the true probabilities since they are consistent. But while they have the same basis and the same asymptotic distribution, the two statistics answer different questions: the null hypothesis of the classic test of (4.1) is that the observations satisfy the given model, the null hypothesis of the LR test of (4.2) is that the given model is as good as the best.

We recall another LR test from Section 3.3 which checks whether the fitted model is an improvement on the null or base-line model with a constant only. This is the test statistic of (3.28),

$$\text{LR}^\circ = 2[\log L(\hat{\beta}) - \log L^\circ],$$

which has k degrees of freedom. With the base-line model we have three loglikelihoods, the base-line model, the fitted model and the saturated model, and the fitted model is as it were suspended between the two extremes. Always omitting the common combinatorial constant, the three likelihoods are, in ascending order,

$$\sum_j n_j \log n_j - n \log n,$$

$$\sum_j [m_j \log \hat{P}_j + l_j (1 - \hat{P}_j)],$$

$$\sum_j (m_j \log m_j + l_j \log l_j - n_j \log n_j).$$

A comparison of the first and second terms tests whether the model

Table 4.1. *Performance tests of car ownership analysis by income class.*

statistic	null	df	value
χ^2	model fits data	$5 - 2 = 3$	2.09
LR*	model equals saturated model	$5 - 2 = 3$	5.70
LR°	income does not matter	$2 - 1 = 1$	17.49**

contributes significantly to an explanation of the variation of the observed frequencies, a comparison of the second and third terms tests for the agreement between the model and the observed frequencies, and a comparison of the first and third terms would test whether the cross-classification of the sample by cells is at all relevant to the incidence of the attribute under review.

We illustrate the tests of this section in Table 4.1 for the only analysis of categorical data at hand, viz. the car ownership example of Section 3.6, with income the sole regressor and five income classes as the cells. At the top of the table we record the goodness-of-fit statistic (4.1); this does not register significance, so that the fitted model is not rejected. There follow two LR tests, constructed from the three loglikelihoods given above, which are -1839.05, -1830.88 and -1828.04 respectively. From the last two values, the test of (4.2) comes out at 5.70, and this is not significant: the fitted model is therefore an acceptable simplification of the saturated model. The LR test of (3.28) is significant, so that income does contribute to an explanation of car ownership. – Altogether the analysis under consideration passes all tests, while it is in fact quite poor. The reason is that the goodness-of-fit tests for categorical data depend on the given classification, and that the classification by five income classes alone is a meagre representation of the distinctions that are relevant to car ownership: with this classification even the saturated model is not much superior to the base-line model.

With a somewhat more sophisticated notation, the arguments and formulae of the present section are easily extended to multinomial models with S alternative outcome states $s = 1, 2, \ldots, S$ instead of just two. They can therefore be readily applied to the models of Chapters 7 and 8.

There is trouble when these goodness-of-fit tests are applied to very large samples. With cell numbers n_j of a thousand or so, the observed frequencies must be very close to the estimated probabilities to fall within the limits of sampling variation. Both goodness-of-fit tests will therefore take significant values, even though the agreement between the f_j and $\hat{P}_j$ looks reasonable enough. It is not certain that this outcome must be taken seriously: the tests allow only for sampling variation and leave no room for the approximate nature of the model specification or for imperfections of the recorded data. Under the null hypothesis, the fitted probability model is a true description of the experiment, and the observations form an impeccable random sample; with increasing sample size the relative frequencies must therefore converge to the prescribed probabilities without fail. In much applied work this is not realistic. It stands in contrast to ordinary regression, where the disturbances continue to reflect imperfections of data and model however large the sample may be.

4.3 The Hosmer–Lemeshow test

The difficulty of designing a likelihood ratio goodness-of-fit test for individual data is the formulation of the ideal model in which the fitted model is nested. The idea of a saturated model with probabilities equal to 1 and a loglikelihood of 0 has played a role in the definition of the *deviance,*

$$D = -\log L$$

as a measure of the (lack of) performance of the fitted model. But such a saturated model does not stand up to asymptotic considerations, for there must be a parameter for each observation to ensure a perfect fit, and the number of parameters will therefore increase beyond all bounds with the sample size. To overcome this problem, Tsiatis (1980) partitions the covariate space into a finite number of J regions, and considers a model with J (0,1) additional dummy variables as the ideal. He then develops a score test of the null hypothesis that these dummy variables have zero coefficients. The test statistic is a quadratic form in the discrepancies

$$f_j - \hat{P}_j$$

that played such a prominent role in the test statistics of the last section. Pregibon (1982) takes up this idea and arrives at a simpler form of the test statistic; he resolves the problem of how to define the J regions (which is not addressed by Tsiatis) by lumping together observations with similar predicted probabilities. The same idea is used in the goodness-of-fit test which Hosmer and Lemeshow developed in a number of papers in the 1980s; see Hosmer and Lemeshow (2000) and the references cited there. We explain this test more fully and apply it to an illustrative example.

The Hosmer–Lemeshow test closely resembles the classic Pearson test of (4.1), with the difference that the cells are not determined by a cross-classification of the covariates, but that the individual observations are ordered into G groups by their estimated probability $\hat{P}_i$. For each group g the expected frequency of successes $\hat{m}_g$ is the sum of the estimated probabilities, and this is compared with the actual frequency m_g. The estimated probability for the group is simply the mean of the $\hat{P}_i$, or $\bar{P}_g = \hat{m}_g/n_g$. The test statistic is

$$C = \sum \frac{(m_g - \hat{m}_g)^2}{n_g \bar{P}_g (1 - \bar{P}_g)},$$

or, in the same form as (4.1),

$$C = \sum n_g \frac{(f_j - \bar{P}_g)^2}{\bar{P}_g (1 - \bar{P}_g)}. \tag{4.3}$$

Under the null hypothesis this has a chi-squared distribution with $G - 2$ degrees of freedom. This test is derived from first principles and supported by a number of simulations in Hosmer and Lemeshow (1980) and further discussed in Lemeshow and Hosmer (1982). But in the asymptotic arguments its authors need the restriction that the number of covariate vectors (or *covariate patterns*) is fixed, or at least that it does not increase in proportion with the sample size. This is tantamount to a cross-classification of the individual observations by categories of the covariates, as in the last section. The assumption is prompted by concern over the behaviour of continuous variables as the sample size increases, with ever increasing numbers of different observed values demanding an ever increasing number of parameters of the perfect model. In the abstract, the number of different values a continuous variable can take is infinite, but in practice this number is limited or at any rate finite. This is certainly so for samples from a real population. In the car ownership data of Section 1.3, for example, urbanization, which is clearly continuous, has been sensibly recorded on a six–point scale. LINC, or the log of income per head, while treated as a genuine continuous variable, is still given with finite precision. Even at absurd levels of precision (which are often employed), the number of different income levels in Holland in 1980, though large, is certainly finite, and for the abstract problem this is enough. In the theoretical exercise there will inevitably come a point when further increases in the sample size will produce more of the same, and from then on there is no need to add new cells to the classification and new parameters to the ideal model.

In the implementation of the Hosmer–Lemeshow test its authors use ten groups, defined either by deciles of $\hat{P}_i$ or by deciles of the ordered sample observations. With ten groups and the range of $\hat{P}_i$ from 0 to 1, the former case gives class limits of $0, 0.1, 0.2, \ldots, 1$, regardless of the actual sample values of $\hat{P}_i$. With unbalanced samples this may lead to a very uneven distribution of the observations over classes, and some classes may have m_g or l_g of less than 5, which is the conventional lower limit. The alternative is to form groups of equal numbers of observations. Lemeshow et al. (1988) compare the two methods for severely unbalanced samples, and advocate caution as the test statistics show somewhat erratic behaviour with small values of m_g and l_g.

The illustration we give bears on a very unbalanced but quite large sample, and it leads to a clear rejection of the fitted model. The example comes from a Dutch bank and refers to over 20 000 bank loans to small business granted in a single year. Two years later some 600 loans are identified as *bad loans*, and this attribute is related to financial ratios of the debtor firm, recorded at the time the loan was granted. Further details are given in Section 6.2. A standard binary logit has been fitted, and this gives the values of m_g and $\hat{m}_g$ shown in Table 4.2. Both panels of this table report the expected and observed numbers of bad loans for ten classes of observations, ordered by $\hat{P}_i$. The first panel employs the standard classification by deciles of $\hat{P}_i$, and we give the overall number of observations by class in the first column. In the second panel the classification is by equal numbers of observations, giving ten classes of 2081 or 2082 observations each; the first column now reports the highest value of $\hat{P}_i$ in each class. The test statistic C of (4.3) is 97.27 for the first classification and 164.35 for the second; both values are highly significant at 8 degrees of freedom, and the logit model is soundly rejected.

The table shows that in the present case the second classification is much more informative than the first, owing to the highly unbalanced nature of the sample. Quite small probabilities prevail, as the mean of

Table 4.2. *Hosmer–Lemeshow test of logit model for bad loans.*

deciles of $\hat{P}_j$			equal sizes		
n_g	$\hat{m}_g$	m_g	$\max\hat{P}_i$	$\hat{m}_g$	m_g
20 453	543	586	0.0100	11	7
233	32	23	0.0158	27	6
60	15	8	0.0198	37	13
26	9	3	0.0229	45	17
11	5	3	0.0258	51	31
9	5	1	0.0286	57	36
11	7	1	0.0317	63	51
4	3	1	0.0359	70	63
6	5	1	0.0443	82	158
3	3	0	0.9148	185	245

predicted probabilities is equal to the sample frequency of 0.03 by the equality of means of Section 3.3. In the classification by deciles of $\hat{P}_i$ observations of either type are therefore very heavily concentrated in the first cell, and the numbers beyond that class are too small to give a clear idea of the quality of the fitted curve. One of the advantages of the Hosmer–Lemeshow test is that it permits inspection of the discrepancies that lead to the rejection of the model: the second classification immediately shows that there is severe overestimation of the probability of bad loans in the lower ranges and underestimation in the two highest classes. This is further illustrated in Figure 4.1, where the expected and actual numbers of bad loans from the second classification have been converted into relative frequencies which are plotted against the argument of the logit function, the log odds ratio of the expected probability,

$$z_g = \log[\bar{P}_g/(1 - \bar{P}_g)].$$

The smooth line of the expected frequencies follows a logistic curve by construction; since the overall sample frequency is so small, it represents the far left-hand tail of the logit function. In contrast, the actual frequencies appear to trace a far larger and much more central part of the same sigmoid shape, if on a reduced scale. This suggests a *bounded* logit

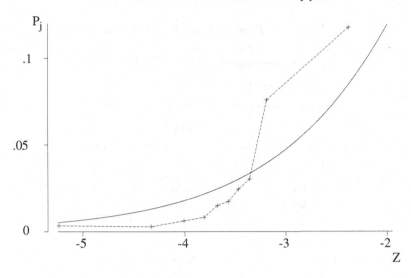

Fig. 4.1. Expected and observed frequencies of bad loans in ten classes.

function with an upper level well below 1, as if a fairly large part of the sample is completely immune to the risk of becoming a bad loan. This variant of the logit model is further discussed in Section 6.2.

4.4 Some measures of fit

The value of the loglikelihood function provides an immediate measure of the model fit; by (3.2) of Section 3.1 the mean loglikelihood, which is often among the standard output of a program package, gives the geometric mean of $\widehat{\Pr}(Y_i)$, the estimated probabilities of the observed outcomes. For the car ownership example of Table 3.4, Section 3.7, this gives $\exp(-1351.39/2820) = 0.61$; but this value largely reflects the overall composition of the sample with an ownership rate of 0.65. Kay and Little (1986) have also proposed direct calculation of the (arithmetic) average of $\widehat{\Pr}(Y_i)$.

Another simple measure of model performance is the *percentage correctly predicted*, also known in marketing research as the *hit rate*. This is found by predicting the outcome by $\hat{P}_i$, employing a cut-off value of 0.5, and counting the matches of predicted and observed outcomes. With unbalanced samples this gives nonsense results: for the bank loan example of the last section, admittedly an extreme case, the numbers are as follows:

$$predicted$$

actual	$Y = 0$	$Y = 1$
$Y = 0$	20 160	623
$Y = 1$	29	4

The percentage correctly predicted is 97%, but it would be the same if we had indiscriminately predicted that all loans are good loans. The result reflects the uneven sample composition, and the prevalent outcome is much better predicted than the rare alternative. This also holds in less extreme cases. The measure is quite sensitive to the sample composition and also to the distribution of $\hat{P}_i$, as argued with some force by (Hosmer and Lemeshow (2000, p. 146–147). More sensible results are obtained by equating the cut-off value to the sample frequency, that is the mean value of $\hat{P}_i$, as I have proposed elsewhere (Cramer 1999). In the present example this is 0.0301, and it gives the following result:

$$predicted$$

actual	$Y = 0$	$Y = 1$
$Y = 0$	13 425	138
$Y = 1$	6764	489

Overall, 67% of the outcomes are correctly predicted, 66% of the $Y_i = 0$ and 78% of the $Y_i = 0$. This is less flattering but more realistic and much more equitable than above. Even so, all such comparisons of individual predictions with the actual outcomes rely on intuitive appeal rather than on a solid theory.

All these measures vary between 0 and 1, and the higher their value the better the performance: in this respect they resemble R^2 of ordinary linear regression. That is a purely descriptive sample statistic which does not contribute to proper statistical inference, yet it has an irresistible intuitive appeal. A large number of R^2-like measures have therefore been put forward for binary discrete models; for reviews see Windmeijer (1995) and Menard (2000). Here we restrict the discussion to two measures that are based on an analysis of variance or decomposition of a sum of squares, just like the classic coefficient of determination of ordinary regression.

We briefly recall the argument from ordinary regression. This starts from the equality

$$\mathbf{y}_r = \hat{\mathbf{y}}_r + \mathbf{e}_r$$

with $\mathbf{y}_r$ and $\hat{\mathbf{y}}_r$ vectors of the observed and estimated values of the dependent variable, and $\mathbf{e}_r$ the residuals; the suffix r indicates that we are dealing with continuous variables in a regression context. Squaring both sides gives

$$\mathbf{y}_r^T \mathbf{y}_r = \hat{\mathbf{y}}_r^T \hat{\mathbf{y}}_r + \mathbf{e}_r^T \mathbf{e}_r,$$

and since $\mathbf{y}_r$ and $\hat{\mathbf{y}}_r$ have the same mean $\bar{Y}_r$

$$\mathbf{y}_r^T \mathbf{y}_r - n\bar{Y}_r^2 = (\hat{\mathbf{y}}_r^T \hat{\mathbf{y}}_r - n\bar{Y}_r^2) + \mathbf{e}_r^T \mathbf{e}_r.$$

The *sum of squares* of $\mathbf{y}_r$, always taken in deviation from the mean, is thus decomposed as

$$\text{SST} = \text{SSR} + \text{SSE}$$

where T, R and E stand for Total, Regression and Error. The proportion of the total variation accounted for by the regression is therefore given by the coefficient of determination

$$R^2 = \frac{\text{SSR}}{\text{SST}} = 1 - \frac{\text{SSE}}{\text{SST}}. \tag{4.4}$$

In addition to the equality of the means of $\mathbf{y}_r$ and $\hat{\mathbf{y}}_r$ this argument makes use of the orthogonality of $\hat{\mathbf{y}}_r$ and $\mathbf{e}_r$,

$$\hat{\mathbf{y}}_r^T \mathbf{e}_r = 0.$$

Since $\mathbf{e}_r$ has zero mean, the elements of $\hat{\mathbf{y}}_r$ can be taken in deviation from the mean, and this is therefore equivalent to $\hat{\mathbf{y}}_r$ and $\mathbf{e}_r$ having covariance zero, or being uncorrelated.

McKelvey and Zavoina (1975) apply the same argument to the latent variable regression equation of (2.8) of Section 2.3. The dependent variable is Y_i°, after normalization; as this is unobserved, SST cannot be directly established. SSR can however be obtained from the variable

$$\hat{Y}_i^\circ = \mathbf{x}_i^T \hat{\boldsymbol{\beta}},$$

while SSE follows from the residual variance, which is set at $\pi^2/3$ for a logit model. This gives the McKelvey-Zavoina measure as

$$R_{\text{MZ}}^2 = \frac{\text{SS of } \mathbf{x}_i^T \hat{\boldsymbol{\beta}}}{\text{SS of } \mathbf{x}_i^T \hat{\boldsymbol{\beta}} + \pi^2/3}.$$

Note that the estimated coefficients are obtained by maximum likelihood estimation of the logit model and not from the latent regression equation under consideration.

The second measure is due to Efron (1978) and it is based directly on the (quasi–)residuals of (3.21) from Section 3.3, which we shall here denote by **e**, omitting the suffix *l*; the argument that follows applies to any probability model, not just to the logit. Efron's R^2 is defined as

$$R_E^2 = 1 - \frac{\mathbf{e}^T \mathbf{e}}{\mathbf{y}^T \mathbf{y} - n\bar{Y}^2}. \tag{4.5}$$

The derivation follows exactly the same lines as for ordinary regression, given above, starting off from the equality of (3.22)

$$\mathbf{y} = \hat{\mathbf{p}} + \mathbf{e},$$

and making use of the same two properties as above, viz. the equality of the means of $\hat{\mathbf{p}}$ and $\mathbf{y}$ and the orthogonality of $\hat{\mathbf{p}}$ and $\mathbf{e}$. For a logit analysis, the equality of the means is assured by (3.25) of Section 3.3, but the orthogonality is not obvious. Efron indeed restricts the discussion to the particular case of categorical variables in order to establish that $\hat{\mathbf{p}}$ and $\mathbf{e}$ are uncorrelated.

Both properties however do hold much more generally, if only asymptotically; for it can be shown that for consistent estimates (like ML estimates) $\hat{\mathbf{p}}$ of *any* probability model the residuals $\mathbf{e} = \mathbf{y} - \hat{\mathbf{p}}$ satisfy

$$\imath^T \mathbf{e}/n \xrightarrow{p} 0 \tag{4.6}$$

and

$$\hat{\mathbf{p}}^T \mathbf{e}/n \xrightarrow{p} 0. \tag{4.7}$$

The first of these is the familiar equality of the means, which happens to hold exactly if the $\hat{\mathbf{p}}$ are ML estimates of a logit model; the second is the *orthogonality* property. A rather laborious proof can be found in Cramer (1997). The asymptotic orthogonality implies that $\hat{\mathbf{p}}$ and $\mathbf{e}$ are approximately uncorrelated in samples of reasonable size; this is easily verified in any particular instance. We shall give some examples below.

The approximate orthogonality vindicates the wider use of Efron's R_E^2, beyond the special case considered by its author. It also has some further implications. From

$$\hat{\mathbf{p}}^T \mathbf{e} \approx 0$$

it follows that

$$\hat{\mathbf{p}}^T \mathbf{y} \approx \hat{\mathbf{p}}^T \hat{\mathbf{p}}$$

so that the sum of the probabilities over the observations with $Y_i = 1$ is equal to the sum of the squared probabilities over the entire sample. Another result, which strengthens the case for Efron's measure, is that it also reflects the discrimination by $\hat{P}_i$ between the two subsamples with $Y_i = 1$ and $Y_i = 0$. To see this we define the mean probabilities in these two subsamples as $\bar{P}^+$ and $\bar{P}^-$. If the sample frequency of success is f, these means must evidently satisfy

$$f\bar{P}^+ + (1-f)\bar{P}^- = \bar{P} = f, \qquad (4.8)$$

and their difference is therefore

$$\bar{P}^+ - \bar{P}^- = \frac{\bar{P}^+ - f}{1 - f}.$$

By (4.8) we may develop $\bar{P}^+$ as

$$\bar{P}^+ = \frac{\hat{\mathbf{p}}^T \mathbf{y}}{nf} \approx \frac{\hat{\mathbf{p}}^T \hat{\mathbf{p}}}{nf} = \frac{\text{SSR} + nf^2}{nf} = \frac{\text{SSR}}{nf} + f.$$

Upon substituting this in the former expression we obtain

$$\bar{P}^+ - \bar{P}^- = \frac{\text{SSR}}{nf(1-f)} = \frac{\text{SSR}}{\text{SST}} = R_{\text{E}}^2. \qquad (4.9)$$

While we cannot give a distribution of R_{E}^2, we can provide a rough idea of what values one may expect to find in the practice of applied work. These values vary with the nature of the data: R_{E}^2 is much higher for analyses of homogeneous samples under the controlled conditions of laboratory trials of bio-assay than for the sample survey data of epidemiology and the social sciences. This is illustrated in Table 4.3 for a number of widely different binary logit analyses.† The first four columns record the subject of the analysis, the sample size n, the number of covariates (including the intercept), and the sample frequency of the most frequent outcome f. The next column gives $\rho_{\hat{p},e}$, the sample correlation between $\hat{\mathbf{p}}$ and $\mathbf{e}$, which should be close to zero by (4.8); and so it is. The last column gives the Efron R^2; the examples have been arranged in ascending order of its value. Apart from the curious steel ingot data, it is sample survey data that lead to quite low values of the order of

† The data of six analyses were kindly provided by their authors and the data of four others were taken from open sources; the intensive care data were copied from the 1989 edition of Hosmer and Lemeshow's textbook.

Table 4.3. *Efron's $R_{\rm E}^2$ for eleven logit analyses.*

	n	$k+1$	f	$\rho_{\hat{p},e}$	$R_{\rm E}^2$
Rolling out of steel ingots, Cox and Snell (1989)	387	3	0.97	0.004	0.044
Depression Los Angeles, 1979, Afifi and Clark (1990)	294	6	0.83	−0.000	0.046
Educational choice of schoolchildren, Holland, 1982, Oosterbeek and Webbink (1995)[a]	1706	12	0.80	0.001	0.072
Antecedents of rapists, Great Britain, 1965–1993, Davies et al. (1997)[a]	210	13	0.84	0.000	0.183
Performance of schoolchildren, Holland, 1965, Dronkers (1993)[a]	699	8	0.78	−0.019	0.208
Employment of women, France, 1979, Gabler et al. (1993)[a]	3658	21	0.52	0.002	0.223
Fibrosis after breast surgery, Holland, 1979–1988, Borger et al. (1994)[a]	332	12	0.72	−0.014	0.260
Private car ownership Holland, 1980	2820	6	0.64	0.009	0.327
Survival in intensive care, Massachusetts, 1983, Lemeshow et al. (1988)	200	9	0.80	0.010	0.359
Toxicity of Tribolium, Hewlett (1969)	1989	4	0.54	0.018	0.385
Hatching mites eggs, Bakker et al. (1993)[a]	149	2	0.58	−0.004	0.504

[a] I thank the authors for kindly making these data sets available to me.

0.1 to 0.3. This is probably comparable to the values of R^2 in ordinary regressions for such data. The best result is obtained for the intensive care example, but this is a nonrandom subset from a larger sample, selected by Hosmer and Lemeshow (2000) for textbook use. The fit is much better (and the samples are more evenly balanced) in traditional laboratory experiments from bio-assay.

5
Outliers, misclassification of outcomes, and omitted variables

This chapter deals with a miscellany of imperfections of data and models. The first subject is the detection of outlier observations, which is largely a matter of common sense and not much different from the approach of ordinary regression. Errors of observation in a discrete dependent variable take the form of misclassification of outcomes, and this is handled by a simple modification of the probability model that has wider applications. The effect of omitted variables on estimation is also different from their role in ordinary regression; this difficult issue has not yet been satisfactorily resolved.

5.1 Detection of outliers

Outliers are observations that do not belong to the sample under review, either because they represent alien elements or because of obvious errors of recording or definition of the variable concerned. It is a good idea to screen the covariate values in the sample, singly or in combination, and to inspect extreme observations for anomalies that may justify their removal. In a narrower definition, outliers are observations that do not obey the relation under review, and these are found by examining the contribution of individual observations to some measure of fit, and singling out those that contribute least (or detract most). The χ^2 statistic of Section 4.2 leads to the squared Studentized or Pearson residual

$$e_i^2 / P_i Q_i,$$

and the Efron R_E^2 of Section 4.4 to the squared (quasi-)residual

$$e_i^2.$$

The simplest measure is $\log L$, and in the representation of (3.2) of Section 3.1 the contribution of observation i is

$$\widehat{\Pr}(Y_i).$$

This is the (estimated) probability of the observed event, and a very small value is a natural indication of an unusual observation.

The squared Pearson residuals have expectation 1, the squared (quasi-) residuals lie between 0 and 1, and so do the probabilities; but this is not enough to set a criterion for what constitutes an extreme value. This must be judged in relation to the other values in the sample under review, bearing in mind the sample size. A practical approach is to rank the sample observations, singling out for further examination extreme values that are separated by a large gap from their neighbours. As $\widehat{\Pr}(Y_i)$ is related to the (quasi-)residual by

$$\widehat{\Pr}(Y_i) = 1 - |e_i|,$$

all three characteristics produce the same ranking and brand the same observations as outliers. In a sample of unequal proportions, the overall level of $\widehat{\Pr}(Y_i)$ will be much lower for the rare outcome than for the predominant alternative, and it is therefore advisable to rank the observations separately for the two outcomes. This means ranking the observations by their contribution to Efron's R_E^2 in its interpretation of (4.9) of Section 4.4.

In the literature the screening for outliers in this sense is often accompanied by the consideration of another diagnostic characteristic which indicates the relevance or technically the *leverage* of each observation. Some observations are more influential in determining the fit or the coefficient estimates than others, and a great deal of sophisticated work has been done to develop characteristics of the individual observations that reflect this property. We refer the reader to Pregibon (1981) and to the treatment in Chapter 5.3 of Hosmer and Lemeshow (2000). Such diagnostic characteristics can serve to bring to light influential data that are not among outliers of the first kind; for once outliers have been identified, their influence is nowadays established without great effort by running analyses with and without the suspected observation. With large samples it would however even now be impracticable to assess the influence of all observations in this fashion.

For a given outlier observation j, we designate the regular sample by the suffix r and the purged or cleaned sample after its removal by c. The change in the loglikelihood upon removing observation j can then

Table 5.1. *Car ownership example: observations with small probabilities.*

	$\widehat{\mathrm{Pr}}(Y_i)$	e_i^2	e_i^2/P_iQ_i
$Y_i = 1$			
817	0.0021	0.9958	465.53
1134	0.0453	0.9115	21.06
17	0.0495	0.9035	19.21
864	0.0509	0.9012	18.67
2221	0.0541	0.8947	17.50
$Y_i = 0$			
752	0.0082	0.9837	121.14
1823	0.0098	0.9805	101.25
2332	0.0251	0.9504	38.85
2694	0.0374	0.9266	25.75
981	0.0421	0.9176	22.74

be partitioned as

$$\log L_{\mathrm{c}} - \log L_{\mathrm{r}} = -\log \widehat{\mathrm{Pr}}_{\mathrm{r}}(Y_j) + \left[\sum_{i \neq j} \log \widehat{\mathrm{Pr}}_{\mathrm{c}}(Y_i) - \sum_{i \neq j} \log \widehat{\mathrm{Pr}}_{\mathrm{r}}(Y_i) \right].$$

Recall that all loglikelihoods are negative, and that the direct effect of deleting an observation is an increase in $\log L$; this is the first term of this partition. The second term is the change in the loglikelihood of the remaining observations brought about by its removal, via the change in the estimated coefficients; this, too, is never negative. It reflects the influence of the observation.

We illustrate these ideas by the car ownership example of Section 3.7. Table 5.1 shows the five smallest values of $\widehat{\mathrm{Pr}}(Y_i)$ for the two subsamples with $Y_i = 1$ and $Y_i = 0$, with the corresponding squared residuals and squared Studentized residuals; the ordering is the same, but the scales are quite different. Observation 817 is the only potential outlier; its value of $\widehat{\mathrm{Pr}}(Y_i)$ is perhaps not too impressive for a sample of almost 3000.†
The effects of its removal are shown in Table 5.2. Neither the coefficient estimates nor the loglikelihoods are much affected. The increase in $\log L$ is 6.2183, and in the partition given above by far the largest part is the

---

† It *is* a powerful outlier in a multinomial analysis of the same data by Windmeijer (1992).

Table 5.2. *Estimates of car ownership model with and without observation 817.*

	regular sample	cleaned sample
LINC	2.38 (16.17)	2.43 (16.39)
LSIZE	2.76 (19.12)	2.80 (19.23)
BUSCAR	-3.04 (19.55)	-3.07 (19.64)
AGE	-0.13 (8.24)	-0.13 (8.18)
URBA	-0.12 (4.20)	-0.12 (4.17)
log L	-1351.39	-1345.18

Absolute values of t − ratios in brackets.

direct effect of $-\log(0.0021) = 6.1658$; there remains only 0.0525 for the second term, which reflects the influence of the omitted observation.

5.2 Misclassification of outcomes

Misclassification of outcomes occurs if Y_i is given as 0 while it should be 1 or the other way around. This can be a recording error, due to simple negligence, but it can also arise from a systematic reporting bias, as when respondents in a sample survey are asked to recall past events from their personal histories. The recorded outcome may also be correct but still at variance with the model under consideration, as in the study of Birnbaum (1968), who wishes to explain the performance of candidates on a multiple-choice test by covariates that represent ability or knowledge. With multiple-choice testing, however, candidates have a fair chance of giving the correct answer even if they do not know what it is. A similar mechanism operates the other way round in the records of a survey where the answer to a sensitive question has been randomized by design.

Probability models can be modified in a simple way to allow for these errors of observation of the dependent variable. Let P° denote the probability of $Y_i = 1$ without misclassification, and suppose that a fixed fraction α_1 of observations with $Y_i = 1$ is erroneously recorded as $Y_i = 0$, with a fraction α_0 for the reverse error. The probabilities of the observed values are

$$P(Y_i = 1) = P_i = (1 - \alpha_1)P_i^\circ + \alpha_0(1 - P_i^\circ),$$
$$P(Y_i = 0) = Q_i = (1 - \alpha_0)(1 - P_i^\circ) + \alpha_1 P_i^\circ. \tag{5.1}$$

This is equivalent to

$$P(Y_i = 1) = P_i = (1 - \alpha_0 - \alpha_1)P_i^\circ + \alpha_0,$$
$$P(Y_i = 0) = Q_i = (1 - \alpha_0 - \alpha_1)(1 - P_i^\circ) + \alpha_1. \tag{5.2}$$

This formulation makes it clear that the modified model is a *bounded* probability model, with the sigmoid curve operating over a reduced range from a lower bound of α_0 to an upper bound of $1 - \alpha_1$. This model is of long standing in bio-assay, where P^* is the probability of death from an insecticide and the two α's represent the forces of natural mortality and immunity of insects, which take precedence over the operation of the poison. Finney (1971) cites an early article by Abbott (1925) advocating this modification of the standard model. In economic applications an upper bound may reflect saturation constraints; there is an upper limit to car ownership because some people cannot drive.

Hausman et al. (1998) made a thorough study of this type of error in responses to survey questionnaires, with an application to past job changes as reported in two large-scale American sample surveys, the Current Population Survey and the Panel Study of Income Dynamics. The authors conclude that there is strong evidence of misclassification at rates of up to 20% or even 30%. These results are obtained by nonparametric estimation, along with standard maximum likelihood estimation of the modified model given above, with P° a probit model. Hsiao and Sun (1999) employ similar constant error rates in the setting of more complex probability models with an application to the market for electronic appliances.

Here we consider the implications of misclassification for the logit. If P° is a standard binary logit probability, the model (5.2) cannot be accommodated by a simple adjustment of the coefficients of the logit transform, and its estimation by maximum likelihood (while not particularly arduous) requires separate programming of the likelihood function and its derivatives.

The effect of ignoring misclassification can be gauged by comparing the results of fitting (5.2) with those of a standard logit. Since P° applies to a restricted range, it is no use comparing the coefficients $\hat{\beta}^\circ$

directly with the estimates of a standard logit, fitted to the same data: we should instead examine the effect of regressors on the probability of the outcome in the two models. The derivative of (5.2), with P° a logit probability, is

$$\partial P / \partial X_j = (1 - \alpha_0 - \alpha_1) P^\circ (1 - P^\circ) \beta_j^\circ.$$

Upon fitting a (misspecified) standard logit $P^\bullet$, the effect is

$$\partial P^\bullet / \partial X_j = P^\bullet (1 - P^\bullet) \beta_j^\bullet.$$

There is a single point where $P = P^\bullet$ and that is the halfway point where both probabilities are 0.5. The slopes are equal at this point if

$$\beta_j^\bullet = (1 - \alpha_0 - \alpha_1) \beta_j^\circ. \tag{5.3}$$

There is some evidence in the tables of Hausman et al., however, that in addition to this adjustment factor there is a misspecification bias that reduces the $\hat{\beta}^\bullet$ further towards zero.

The operation of this mechanism is illustrated for a simple case in Figure 5.1. The solid line represents the course of P of (5.2) with $\alpha_0 = \alpha_1 = 0.2$ and $0 + 1.5X$ as the argument of the logit P°. The dotted line is $\hat{P}^\bullet$ from a misspecified ordinary logit fitted to simulated values of P. These simulations mimic the latent regression equation (2.9) of Section 2.3. The single covariate X is a standard normal variate, established once and for all for a sample of size 3000; in each of 100 replications the latent regression is completed by drawing fresh disturbances from a logistic distribution with mean zero and variance $\pi^2/3 \approx 3.29$. The Y_i

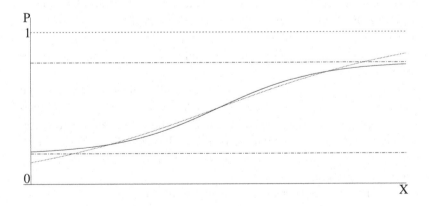

Fig. 5.1. Logit curve with errors of observation.

Table 5.3. *Bias through ignoring misclassification.*

	$(1 - 2\alpha)\beta°$	mean $\hat{\beta}^{\bullet}$	ratio
$\alpha = 0.02$	1.44	1.39 (.006)	0.96
$\alpha = 0.05$	1.35	1.24 (.005)	0.92
$\alpha = 0.10$	1.20	1.04 (.005)	0.87
$\alpha = 0.20$	0.90	0.72 (.004)	0.80

Standard errors from replications in brackets.

are determined from the usual inequality, and then changed at random from 1 to 0 and from 0 to 1 by the equal misclassification probabilities $\alpha_0 = \alpha_1 = \alpha = 0.1$. An ordinary logit is then fitted by a standard routine to these partly misclassified outcomes. By (5.3) the slope coefficient should be $0.6 \times 1.5 = 0.9$ but the mean over the replications is only 0.72, and this value has been used to draw the dotted line. The misspecified logit function is apparently distorted by fitting it to the constrained range so as to give a weaker slope. We have repeated this exercise for various values of α; Table 5.3 shows that the bias due to neglect of the misclassification increases with its extent, as was to be expected.

In the form of (5.2) the present modification is easily generalized to multinomial models, with separate α_s for each state s, but I know of no empirical applications.

5.3 The effect of omitted variables

When a treatment is tested on a homogeneous sample under controlled laboratory conditions, the effect of other covariates can be safely ignored. In randomized field trials and the surveys of epidemiology and the social sciences, however, there are always contingent regressor variables at work, and there are always some missing from the analysis through oversight or lack of data. Marketing and finance analyses may employ twenty or thirty regressors, but even then some determinants will be absent and lead to unobserved heterogeneity. Ordinary regression is little affected by these imperfections: provided the omitted variables are not correlated with the remaining regressors, the coefficient estimates are still consistent and unbiased, and the only inconvenience is a loss of precision due to increased residual variance. No such comforting argument holds for the logit and probit models. We shall see that even if the omitted variables are uncorrelated with the remaining covariates they will affect the parameter estimates, usually (but not always) biasing them towards zero.

There are some scattered references to omitted variables in the literature. Gail et al. (1984) consider a simple case–control setting with binary outcome, binary treatment and a binary additional covariate, and show that its neglect leads to a bias towards zero in two nonlinear models, namely the logistic regression and hazard models. In econometrics, a number of authors have attempted (and failed) to find an analogy to the omitted variable paradigm of ordinary regression: see Lee (1982), Ruud (1983) or Gourieroux (2000). Amemiya and Nold (1975) allow for omitted variables by adding an extra disturbance to the logit transform of categorical data.

The present treatment is simpler than these sophisticated theoretical studies. We trace the effect of an omitted (but relevant) variable through the latent variable equation (2.8) of Section 2.3

$$Y_i^* = x_i^T \beta^* + \varepsilon_i^*.$$

This has all the standard properties of ordinary regression; ε^* has zero mean and variance σ^{*2} and is uncorrelated with the regressors. The sign of the latent variable Y_i^* determines the $(0,1)$ observed indicator variable Y_i. We recall that β^* and σ^{*2} are not identified, and that this is resolved by setting the value of σ^{*2} at some *a priori* value C^2 and estimating the *normalized* coefficients

$$\beta = \frac{C}{\sigma^*} \beta^*.$$

For the logit model the constant C is $\pi/\sqrt{3} \approx 1.814$.

The effect of omitting a relevant variable is examined by considering the removal of X_2 from a *reference equation* with two regressors

$$Y_i^* = \beta_0^* + \beta_1^* X_{1i} + \beta_2^* X_{2i} + \varepsilon_i^*, \qquad (5.4)$$

which will give a *curtailed* equation with only one covariate. X_2 may be projected on to the unit constant 1 and X_1 as in

$$X_{2i} = \gamma_0^* + \gamma_1^* X_{1i} + v_i^*, \qquad (5.5)$$

which again has the same properties as an ordinary regression equation; the variance of v^* is σ_v^2. Substitution gives the curtailed equation as

$$Y_i^* = (\beta_0^* + \beta_2^*\gamma_0^*) + (\beta_1^* + \beta_2^*\gamma_1^*)X_{1i} + \varepsilon_i^\circ \qquad (5.6)$$

with

$$\varepsilon_i^\circ = \varepsilon_i^* + \beta_2^* v_i^*.$$

We disregard the change in the intercept, which had already absorbed a nonzero mean of ε^* and now (with γ_0^*) absorbs a nonzero mean of v^* as well. The main effects of omitting X_2 are a direct change in the coefficient of X_1, from β_1^* to $\beta_1^* + \beta_2^*\gamma_1^*$, and an increase of the residual variance from σ^{*2} to $\sigma^{\circ 2} = \sigma^{*2} + \beta_2^{*2}\sigma_v^2$. As a result, the *normalized* coefficient of X_1 in the curtailed equation is

$$\beta_1^\circ = \frac{C}{\sqrt{\sigma^{*2} + \beta_2^{*2}\sigma_v^2}} (\beta_1^* + \gamma_1^*\beta_2^*). \qquad (5.7)$$

The interesting point is that even if X_2 is not related to X_1 its removal still affects the remaining coefficient. The regressors are uncorrelated, or orthogonal, or there is no confounding among the covariates if $\gamma^* = 0$, and in this case the projection of (5.5) collapses to

$$X_{2i} = \gamma_0^* + v_i^* = \bar{X}_2 + v_i^*$$

where $\bar{X}_2$ is the sample mean of X_2. As a result, $v_i^* = X_{2i} - \bar{X}_2$ and $\sigma_v^2 = \text{var } X_2$. The curtailed equation (5.6) now becomes

$$Y_i^* = (\beta_0^* + \beta_2^*\bar{X}_2) + \beta_1^* X_{1i} + \varepsilon_i^\circ \qquad (5.8)$$

with

$$\text{var } \varepsilon_i^\circ = \sigma^{\circ 2} = \sigma^{*2} + \beta_2^{*2}\text{var } X_2. \qquad (5.9)$$

The coefficient of X_1 in the latent variable equation is not affected by the removal of X_2, but the *normalized* coefficient declines because of the increase in the residual variance. In terms of β_2 instead of β_2^* we find

$$\frac{\beta_1^\circ}{\beta_1} = \frac{1}{\sqrt{1 + \beta_2^2\text{var } X_2/C^2}}. \qquad (5.10)$$

The same ratio holds for all slope coefficients if there are several left. Omitting orthogonal variables thus leads to a systematic tendency or bias towards zero of the remaining coefficients. The ratio (5.10) is the *rescaling factor* of Yatchev and Griliches (1985); its size depends on the *impact* of the omitted variable, relative to the imposed variance C^2, that is on

$$\text{impact} = \beta_j^2 \text{var} X_j. \tag{5.11}$$

Since β_j can be estimated in the reference model, this argument can be empirically verified, provided we can find an orthogonal regressor to delete.

The addition of $\beta_j X_{ji}$ to the disturbances will also affect the shape of their distribution. X_j must have a very special sample distribution indeed for both ε^* and ε° to satisfy a logistic distribution; it must be distributed like the difference of two independent logistic variates.† In the literature this point is often overlooked, though not by Ford et al. (1995) for hazard models, where it is particularly relevant. Apart from this special case, the two models cannot be simultaneously valid, and one or both is misspecified. We shall see that this may lead to further systematic changes in the estimated coefficients. It stands to reason that the size of this misspecification effect will vary along with the impact; but in contrast to the rescaling effect, its direction is uncertain.

As an illustration consider car ownership as a function of income and habitat, which is a simple dichotomy by town and country. Within each group car ownership rates vary with income along logit curves with the same coefficient β_1 of log income but different intercepts β_0; this shift of the ownership Engel curves reflects the fact that at equal incomes country people have a higher rate of car ownership than city dwellers. In Figure 5.2 the two separate logit curves are shown as solid lines. If the distinction between the two strata is ignored and they each provide half the sample, overall car ownership varies with income as the average of the two logistic functions, shown by the dotted line. At income level X_1° overall car ownership is 0.5 and the specific ownership rates must add up to 1; we put them at α for the city and $1 - \alpha$ for rural areas. The slope of the two logistic curves is $P Q \beta_1$, and at this point it is the same for both strata and equal to $\alpha (1 - \alpha) \beta_1$. The same slope must hold for the average curve; but while this is sigmoid in shape, it is not a logit curve, for the average of two logit curves is not a logit (nor is the average of two probit curves a probit curve). If nevertheless a logit is fitted to the overall sample its slope at X_1° will be $0.5 \times 0.5 \beta_1^\circ$. This must be approximately equal to the slopes of the separate curves, and

---

† In a probit model X_j must have a normal distribution to meet this requirement; most people feel more comfortable with this, but it is equally restrictive.

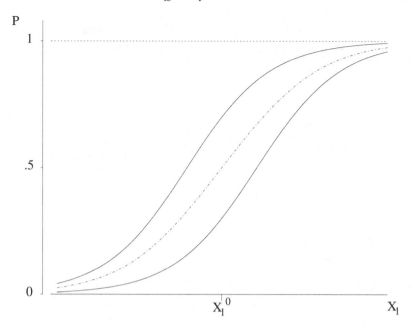

Fig. 5.2. Logit curves for car ownership in two strata and overall.

β_1° will be lower than β_1 by a factor

$$\frac{\beta_1^\circ}{\beta_1} = \frac{\alpha(1 - \alpha)}{0.25}.$$

Ignoring the strata thus leads to a slope coefficient that is closer to zero.

The same example can be treated by numerical simulation. The two separate curves correspond to a reference latent regression with two variables,

$$Y_i^* = \beta_0 + \beta_1 X_{1i} + \beta_2 X_{2i}.$$

X_1 is the log of income and a X_2 a binary habitat dummy taking the values 0, 1. The curtailed regression has X_1 alone. We have constructed a sample of size 3000 with X_1 a standard normal variate and X_2 a $(0, 1)$ dummy variable with a 50/50 distribution; the two are independent. In the reference equation β_1 is 1.8; β_0 and β_2 vary with the shift parameter α that was introduced above. The latent variable regression equation is completed by drawing disturbances from a logistic distribution with mean zero and variance $\pi^2/3 \approx 3.29$, so that no normalization of the coefficients is needed. The Y_i are determined by the condition $Y_i^* > 0$.

Table 5.4. *Effect of omitted variables from simulations.*

simulation	impact factor	rescaling factor	mean simulated coefficient ratio
strata dummy			
α=0.40	0.1645	0.9759	0.9634 (0.00077)
α=0.35	0.3833	0.9464	0.9263 (0.00093)
α=0.30	0.7182	0.9060	0.8568 (0.00140)
α=0.25	1.2074	0.8553	0.7886 (0.00135)
normal X_2			
1	0.1674	0.9755	0.9725 (0.00064)
2	0.4003	0.9442	0.9302 (0.00091)
3	0.7285	0.9048	0.8918 (0.00112)
4	1.1746	0.8584	0.8138 (0.00133)

Standard errors from replications in brackets.

Both the reference model and the curtailed model are fitted by standard maximum likelihood methods and the ratio of the two estimates of β_1 is established. This exercise has been repeated in 100 replications with the same X_1 and X_2 but fresh disturbances.

The results are reported in the top panel of Table 5.4. In the first line $\alpha = 0.4$, and this gives an impact (5.11) of 0.1645 and a rescaling factor of 0.9759. The simulations however give a mean coefficient ratio of 0.9634 with a standard deviation of 0.00077: there is an additional downward shift that must be attributed to the misspecification of the distribution. The overall effect is still small, but then the impact is quite small relative to C^2 of 3.29. In the next three simulations α moves away from 0.5, the gap between the strata widens, the impact and the rescaling factor increase, and so does the misspecification effect.

We have added simulations where X_2 is a normal variate instead of a binary dummy. With roughly similar impacts and rescaling factors the misspecification effect is much smaller. Clearly this effect depends on the distribution of the omitted variable; it appears that it is particularly sensitive to its kurtosis. The kurtosis of the binary dummy is negative, and the kurtosis of the normal variate is zero; these values stand in contrast with the kurtosis of 1.2 of the logistic distribution with its fat tails. Further experiments have shown that for X_2 with a positive kurtosis the misspecification effect disappears; for high values it may even change sign, and partly offset the rescaling effect.

The conclusion is that omitted variables do affect the estimated coefficients of the remaining covariates, even if they are orthogonal, and that in this case they are likely to depress the estimates. The rescaling effect leads to a downward bias, and it is accompanied by a misspecification effect that depends on the distribution of the omitted variable. This will often strengthen the rescaling bias, but in the final analysis its size and sign are uncertain.

In practice, we do not know the omitted variables, we can only speculate about their effect, and we cannot remedy it – the more so as we cannot be sure of the logistic distribution of the disturbances of the reference equation in the first place. The only practical conclusions are that it is important to include all relevant covariates in the analysis, and that meta-analyses must be approached with caution. Estimates from different studies are seldom comparable as the omitted variables are bound to vary from one analysis to another.

All this refers to the estimated slope coefficients. As we have explained in Section 2.4, the purpose of the analysis often goes beyond their calculation. In their further use for purposes of selection or discrimination the omitted variable effect is mitigated, for if all estimated coefficients are biased towards zero in the same proportion, the rank order of the observations by $\hat{P}_i$ remains the same and so does the selection; but the cut-off criterion may have to be adjusted. The use of estimated probabilities in conditional aggregate predictions may suffer more from the omitted variable effect on the coefficients. In this application, however, it is much more important whether it is justified to ignore the omitted variables in the larger sense that they are assumed constant in the *ceteris paribus* condition of the prediction.

Finally, we give an empirical demonstration of the effect of deleting variables from a logit analysis for the car ownership example of Section 3.7. The final equation with five regressors of Table 3.4 is the reference equation, and four regressors are removed one by one until only LINC remains. The order of this process is suggested by the statistics of

Table 5.5. *Household car ownership: statistics of regressor variables.*

variable	LINC	LSIZE	BUSCAR	AGE	URBA	impact	rescaling factor
LINC	1	−0.56	−0.01	0.06	0.14	1.23	0.85
LSIZE		1	0.15	−0.22	−0.22	1.84	0.80
BUSCAR			1	−0.11	−0.08	0.98	0.88
AGE				1	0.03	0.16	0.98
URBA					1	0.04	0.99

Table 5.5. The last three variables are almost uncorrelated with LINC and therefore the first candidates for removal, even though their impacts and rescaling factors are quite small. Cursory inspection of a handful of similar studies shows that the present example provides a better illustration than most: quite often *all* explanatory variables have negligible impacts, even though the estimated coefficients are significantly different from zero. This may indicate that the reference equation itself is

Table 5.6. *Household car ownership: effect of removing regressor*
variables on remaining coefficients.

	all five	less URBA	less AGE	less BUSCAR	less LSIZE
LINC	2.38 (0.15)	2.36 (0.15)	2.46 (0.15)	1.77 (0.13)	0.35 (0.09)
LSIZE	2.76 (0.14)	2.83 (0.14)	3.09 (0.14)	2.22 (0.12)	
BUSCAR	−3.04 (0.16)	−3.00 (0.15)	−2.95 (0.16)		
AGE	−0.13 (0.02)	−0.12 (0.02)			
URBA	−.12 (0.03)				

Standard errors in brackets.

severely incomplete, so that omitted variable bias has already depressed the reference estimates at the start.

The effect of the successive removal of all regressors except LINC is shown in Table 5.6. URBA and AGE have so little impact that they can be omitted without affecting the other coefficients, as can be seen by comparing the first three columns. But BUSCAR, while only faintly correlated with LSIZE and LINC, is a powerful explanatory variable, and upon its removal the remaining two coefficients are reduced to 0.72 of their previous values. This is far more than the rescaling effect of 0.80, but we know from the simulations that the removal of a binary variable has a strong downward misspecification effect. The last remaining regressor, LSIZE, has a strong negative correlation with LINC, and we must therefore use (5.6) to assess the effect of its removal. A regression of LSIZE on LINC gives $\hat{\gamma}_1^* = -0.59$, and this goes a long way to explain the sharp decline of the income coefficient.

6

Analyses of separate samples

In Section 3.2 we have described the data as resulting from a series of laboratory experiments or from sampling a real population. In either case they are generated by a single process and form a single entity, even though the elements differ in outcome and covariates. This view is now abandoned and the data are distinguished by outcome and treated as separate groups, or even collected as two distinct samples. In discriminant analysis the data are regarded as a mixed sample from two different populations. We next consider trimming a single sample by discarding observations with the more numerous outcome, which is equivalent to drawing separate samples for each outcome. Finally we briefly examine case-control studies, which are the ultimate example of using separate samples.

At first sight it must make a difference whether the data are a single sample from a mixed population or two separate samples from different outcome groups. The logit model surprisingly applies in either case.

6.1 A link with discriminant analysis

Discriminant analysis is a statistical technique for classification and selection. In its simplest form (which is the only form considered here) it starts off from the assumption that the sample observations are drawn in proportions λ and $1 - \lambda$ from two populations, groups or classes, labelled 1 and 0. The elements of the two groups differ systematically in k characteristics $\tilde{\mathbf{x}}_i$, a vector of proper covariates without a unit constant. In each of the two populations the $\tilde{\mathbf{x}}_i$ have a k-dimensional normal distribution with the same covariance matrix Σ but with different means $\boldsymbol{\mu}_1$ and $\boldsymbol{\mu}_0$. An element with characteristics $\tilde{\mathbf{x}}_o$ is observed, and we wish to assign it to one of the two groups. In practice, this question arises when components like nuts or bolts are produced by two machines, or bought from two suppliers, and the origin of a defective item is in question. Botanists may wish to assign plants or flowers to a particular species, and entomologists to classify insects without clear sex characteristics as male or female. In finance the technique is used to identify firms that are likely to go bankrupt or that are targets of take-over bids, and in marketing it serves to classify consumers into market segments with different preferences.

In discriminant analysis, this statistical problem is solved by finding a linear function Z of $\tilde{\mathbf{x}}$ that best separates the two groups. One criterion for doing so is to maximize the ratio of between-group variance to within-group variance, taking into account the (relative) costs of the two kinds of potential misclassification as well as the proportion λ. But we may equally well follow a roundabout route and solve the problem in two steps by first finding the probability that an element belongs to the target group and then applying a cut-off criterion as discussed in Section 2.4. This route leads to the logit model and we shall adopt it here, following the derivation of Ladd (1966); more about discriminant analysis can be found in the monograph of Lachenbruch (1975).

The probability that an item with characteristics $\tilde{\mathbf{x}}_\circ$ belongs to group 1 is established by Bayes' theorem. For two events A and B we have

$$P(AB) = P(A)P(B|A) = P(B)P(A|B)$$

and hence

$$P(A|B) = \frac{P(A)P(B|A)}{P(B)}. \tag{6.1}$$

For the present problem, we define

- A as the event "the observed item belongs to group 1";
- $\bar{A}$ as its complement, "the observed item belongs to group 0";
- B as the event "the observed item has characteristics $\tilde{\mathbf{x}}_\circ$".

The probabilities are given by

$$P(A) = \lambda,$$
$$P(\bar{A}) = 1 - \lambda,$$
$$P(B|A) = \phi_1(\tilde{\mathbf{x}}_\circ),$$
$$P(B|\bar{A}) = \phi_0(\tilde{\mathbf{x}}_\circ),$$

where ϕ_1 and ϕ_0 denote the k-dimensional densities of $\tilde{\mathbf{x}}$ in the two populations. It follows that

$$P(B) = P(B|A)P(A) + P(B|\bar{A})P(\bar{A})$$
$$= \lambda\phi_1(\tilde{\mathbf{x}}_o) + (1 - \lambda)\phi_0(\tilde{\mathbf{x}}_o).$$

Substitution of these probabilities into (6.1) gives the probability that an item with $\tilde{\mathbf{x}}_o$ belongs to group 1, or, with an indicator function Y for this event,

$$P(Y = 1|\tilde{\mathbf{x}}_o) = P(A|B) = \frac{\lambda\phi_1(\tilde{\mathbf{x}}_o)}{\lambda\phi_1(\tilde{\mathbf{x}}_o) + (1 - \lambda)\phi_0(\tilde{\mathbf{x}}_o)}. \qquad (6.2)$$

The corresponding odds are

$$O(Y = 1|\mathbf{x}_o) = \frac{\lambda\phi_1(\tilde{\mathbf{x}}_o)}{(1 - \lambda)\phi_0(\tilde{\mathbf{x}}_o)}$$

and the log odds or logit is

$$R(Y = 1|\tilde{\mathbf{x}}_o) = \log[\lambda/(1 - \lambda)] + \log\phi_1(\tilde{\mathbf{x}}_o) - \log\phi_0(\tilde{\mathbf{x}}_o).$$

We now make use of the specification of ϕ_1 and ϕ_0 as multinomial normal densities,

$$\phi_1(\tilde{\mathbf{x}}_o) = C \exp -\frac{1}{2}[(\tilde{\mathbf{x}}_o - \boldsymbol{\mu}_1)^T\boldsymbol{\Sigma}^{-1}\tilde{\mathbf{x}}_o - \boldsymbol{\mu}_1)],$$

$$\phi_0(\tilde{\mathbf{x}}_o) = C \exp -\frac{1}{2}[(\tilde{\mathbf{x}}_o - \boldsymbol{\mu}_0)^T\boldsymbol{\Sigma}^{-1}(\tilde{\mathbf{x}}_o - \boldsymbol{\mu}_0)],$$

with C the constant $\pi^{k/2}|\boldsymbol{\Sigma}|^{-1/2}$. Substitution into the preceding formula gives

$$R(Y = 1) = \log[\lambda/(1 - \lambda)] + \frac{1}{2}[(\tilde{\mathbf{x}}_o - \boldsymbol{\mu}_1)^T\boldsymbol{\Sigma}^{-1}(\tilde{\mathbf{x}}_o - \boldsymbol{\mu}_1)]$$
$$- \frac{1}{2}[(\tilde{\mathbf{x}}_o - \boldsymbol{\mu}_0)^T\boldsymbol{\Sigma}^{-1}(\tilde{\mathbf{x}}_o - \boldsymbol{\mu}_0)].$$

This can be simplified by expanding and rearranging terms; note that terms like $\tilde{\mathbf{x}}_o\boldsymbol{\Sigma}^{-1}\boldsymbol{\mu}_1$ are scalars, and hence invariant under transposition. The end result is

$$R(Y = 1) = \log[\lambda/(1 - \lambda)] - \frac{1}{2}[(\boldsymbol{\mu}_1 + \boldsymbol{\mu}_0)^T\boldsymbol{\Sigma}^{-1}(\boldsymbol{\mu}_1 - \boldsymbol{\mu}_0)]$$
$$+ \tilde{\mathbf{x}}_o^T\boldsymbol{\Sigma}^{-1}(\boldsymbol{\mu}_1 - \boldsymbol{\mu}_0). \qquad (6.3)$$

The log odds or logit of $P(Y = 1)$ is therefore the sum of a constant and a linear function in $\tilde{\mathbf{x}}_o$; in other words, $P(Y = 1)$ follows a standard logit model with argument $\mathbf{x}_o^T\boldsymbol{\beta}$. Upon partitioning $\mathbf{x}$ and $\boldsymbol{\beta}$ as in (2.11), the intercept β_0 is given by the first line of (6.3), and the slope coefficients

$\tilde{\beta}$ by the second line. The population proportion λ (which may be unknown) enters only in the intercept; the slope coefficients reflect the contrast between the two populations as expressed by $\mu_1 - \mu_0$.

Once the logit function has been thus obtained its origins may be forgotten, and it can be estimated by the standard routines without regard to the specification of a normal distribution of the regressors. This will yield estimated probabilities for any x_o, and the item concerned can then be allotted to either population by a *cut-off* criterion, derived by minimizing the expected loss from misclassification, as has been sketched in Section 2.4.

From the viewpoint of discriminant analysis this is a roundabout route, and it is much more sensible to estimate the optimal discriminant function directly, taking into account the proportion λ and incorporating misclassification costs into the estimation procedure. This does not give the same estimates as logistic regression, and the resulting classification of elements may also be different, especially if the assumption of a normal distribution of the covariates is poorly satisfied. There is an extensive literature on the relative performance of the two methods; for recent examples from finance see Lennox (1999) and Barnes (2000).

In further developments of the method, the rigid assumptions listed above can be relaxed by allowing for different covariance matrices for the groups, or by considering other distributions of the regressors than the normal. These alternatives may still yield logit functions for $P(Y = 1)$, in some cases with transformations of the regressor variables in the argument. Other distributions of the covariates can also lead to different probability functions; see Kay and Little (1987). But while these alternatives widen the field they still impose strict (and somewhat contrived) conditions on the distribution of the covariates.

Here, discriminant analysis is brought in only to demonstrate how the logit function may arise from a different model with a quite different background. Discriminant analysis is a more natural approach to classification or selection than the logit model; the notion that observations belong to distinct populations and that their allegiance can be inferred from their covariates is alien to its strong causal tradition. Technically, the two models differ in the importance of distributional assumptions about the covariates: even with departures from normality this remains an important issue in discriminant analysis, while it is of no concern in logistic regression. These viewpoints cannot be reconciled, and the link between the two models is largely a mere accident of algebra. Still it can occasionally be instructive to consider the logit in the light of discriminant analysis, as we shall see in the next section.

6.2 One-sided sample reduction

One-sided sample reduction is the simplest form of endogenous sample selection. This practice must surely have arisen before its theory was developed, for it is the only easy way of analysing rare attributes in very large samples. A direct mail firm sends circular letters to hundreds of thousands of potential customers with a few known covariates (like zip code and ordering record), and receives only a few hundred orders in return. A statistical analysis of this response is in order. In view of the vast numbers involved the analyst may decide to contrast the small group of respondents with an equal number of nonrespondents drawn at random. We shall see that this can be a sensible and valid procedure. It was originally prompted by technical obstacles to data handling and computations with vast numbers of observations, but these no longer hold. There is however still a strong case for restricting the sample size if additional data must be collected. In the analysis of take-over targets of Palepu (1986), the initial sample consist of all companies that are quoted on the stock exchange; only a tiny fraction has been subjected to a take-over bid. Since further information on the companies in the sample must be collected by documentary research, it will pay to consider all subjects of take-over bids but only part of the others.

Suppose the *initial* or *full* sample consists of a small number m of observations with state 1 and a very large number l with state 0. Upon taking only a fraction λ of the latter, drawn at random, we obtain a *reduced sample* of m observations with state 1 and $l^{\bullet} = \lambda\, l$ observations with state 0. The probability model

$$P(Y_i = 1|\mathbf{x}_i) = P_i^* = P^*(\mathbf{x}_i, \boldsymbol{\theta}) \tag{6.4}$$

applies to all observations of the full sample, and hence also to the elements of the reduced sample. But the basic conditions of maximum likelihood theory as set out in Section 3.2 no longer apply, and in particular it is no longer legitimate to invoke 'conditioning on the covariates'. The data may still be submitted to a routine maximum likelihood estimation, but the resulting estimates will not possess the optimal properties listed at the end of Section 3.1 that form the main attraction of this method.

One way of facing this complication is to relate the probability of $Y = 1$ for an element of the reduced sample explicitly to its probability in the full sample, and to adjust the likelihood accordingly. This is done again by an application of Bayes' theorem of (6.1),

$$P(A|B) = \frac{P(A)P(B|A)}{P(B)}.$$

This time we define
- A as the event "$Y_i = 1$";
- $\bar{A}$ as its complement, "$Y_i = 0$";
- B as the event "element i is part of the reduced sample".

The probabilities are

$$P(A) = P_i^*,$$
$$P(\bar{A}) = 1 - P_i^*,$$
$$P(B|A) = 1,$$
$$P(B|\bar{A}) = \lambda.$$

This gives

$$P(B) = P(B|A)P(A) + P(B|\bar{A})P(\bar{A})$$
$$= P_i^* + \lambda(1 - P_i^*),$$

so that the probability of $Y = 1$ for an element of the reduced sample is

$$P_i = \frac{P_i^*}{P_i^* + \lambda(1 - P_i^*)}. \tag{6.5}$$

λ is a known constant, set at will by the analyst; upon writing P^* in full as $P^*(\mathbf{x}_i, \boldsymbol{\theta})$, P_i is a function of $\boldsymbol{\theta}$ alone. The loglikelihood of the reduced sample is therefore easily expressed as a function of $\boldsymbol{\theta}$, too, and these parameters can be estimated from the reduced sample by standard maximum likelihood methods, though as a rule not by a standard routine.

In the special case that (6.4) is a logit probability, estimation is much easier. For the odds of (6.5) we have

$$\frac{P_i}{1 - P_i} = \frac{P_i^*}{\lambda(1 - P_i^*)}$$

and for the log odds or logit

$$\text{logit}(P_i) = -\log \lambda + \text{logit}(P_i^*). \tag{6.6}$$

If P^* is a binary logit probability, $\text{logit}(P_i^*)$ is a linear function of $\mathbf{x}_i$, namely $\mathbf{x}_i^T \boldsymbol{\beta}$; by (6.6) $\text{logit}(\tilde{P}_i)$ is almost the same linear function, with the intercept adjusted to $\beta_0 - \log \lambda$ but the same slope coefficients $\tilde{\boldsymbol{\beta}}$.

Thus the P_i of the selected sample also obey a standard logit model with the same slope coefficients as the logit model for the full sample. This powerful result, first noticed by Prentice and Pyke (1979), sets the logit specification apart from all other models; its major implications are sketched in the next sections.

It follows that standard logistic regression routines may be applied without further ado to the reduced sample to give proper estimates of the slope coefficients; if needs be, the full sample intercept can be retrieved by adding $\log \lambda$ to the intercept of the reduced sample. For purposes of selection the intercept is however immaterial, as the ordering of elements by their estimated probability remains the same if the intercept varies (though the cut-off point must be adjusted). The selection of take-over targets is a case in point, as the main purpose is to identify a few companies most likely to be the subject of a bid.

We illustrate this technique by two examples. The first, from Cramer et al. (1999), concerns a Dutch financial institution which offers a variety of savings accounts and investment funds. Its clients can easily shift their holdings from one product to another. For almost six years monthly shifts in investments have been recorded for 9600 clients with a positive balance in a savings account and no other investment at the beginning of the month. The issue is whether they switch part of their holdings to other products (and thereby drop from the sample). The data consist of 293 880 monthly observations with only 1488 switches, or a sample frequency of 0.0051. In a preliminary screening of the material a simple binary logit is fitted with five covariates. Two are customer characteristics: LOYALTY, or the length of time a client has been with the firm, and SAVINGS, the amount in the savings account (in logarithms). Three are outside events during the month, namely SHARE INDEX, last month's return; INTEREST RATE, change from previous month; and NEW PRODUCT, a binary dummy for the introduction of a new product.

We compare the logit estimates of the full sample with $l = 292\,392$ and of a reduced sample with an equal number of zero observations, drawn at random, or $l^\bullet = 1488.$† The result is shown in Table 6.1; the intercept of the reduced sample has been re-adjusted by adding $\log \lambda$. The estimates from the two samples show a remarkable resemblance, even if is borne in mind that they have the observations with $Y = 1$ in common. The only appreciable difference occurs in the coefficients of

† This was done by the standard maximum likelihood routine of GAUSS; no special provisions were needed beyond an increase of the program memory and an upgrade of the working memory of the desk computer to 320 MB.

Table 6.1. *Switching savings to investments:*
estimates from three samples.

λ	0.0051	1	0.0254
1	1488	292 392	7440
intercept	−4.85 (0.12)	−4.87 (0.08)	−4.86 (0.09)
LOYALTY	−.29 (0.10)	−.26 (0.07)	−.21 (0.08)
SAVINGS	.47 (0.04)	.46 (0.02)	.46 (0.03)
SHARE INDEX	1.39 (0.53)	1.51 (0.35)	1.18 (0.39)
INTEREST RATE	−1.20 (0.16)	−1.21 (0.12)	−1.14 (0.13)
NEW PRODUCT	0.85 (0.08)	0.76 (0.05)	0.76 (0.06)

Standard errors in brackets.

SHARE INDEX, which are less precise than the others, as their standard errors show.

The second example is the incidence of bad loans at a Dutch bank. This bank granted over 20 000 bank loans to small business in a single year; two years later some 600 loans turned out to be *bad loans*. This does not mean that the debtors default: a bad loan is a loan that causes trouble and demands the attention and time of bank managers. In addition to this dichotomy the data set reports six covariates, all financial ratios of the debtor firm, recorded when the loan was granted; five are used in a logit analysis, viz.

- SOLVENCY or the ratio of own capital to the balance sheet total;
- RENTABILITY or the ratio of gross returns to the balance sheet total;
- WORKING CAPITAL or the ratio of working capital to the balance sheet total;
- CASH FLOW COVERAGE or the ratio of cash flow to interest due;
- STOCKS, a $(0,1)$ dummy variable for the presence of material stocks.

As these financial ratios have been taken mechanically from the firms' accounts, extreme values occur easily when the denominator is close to zero; out of 20 862 observations 46 were discarded as outliers because of one or more extreme regressor values, separated by substantial gaps

Table 6.2. *Incidence of bad loans: estimates from*
reduced and full sample.

	mean of 100 estimates from reduced samples	full sample
SOLVENCY	1.65 (0.34)	−0.49 (0.09)
RENTABILITY	−0.93 (0.25)	−0.43 (0.12)
WORKING CAPITAL	−1.41 (0.28)	−0.90 (0.12)
CASH FLOW COV.	−2.96 (1.52)	−2.60 (0.40)
STOCKS	0.13 (0.20)	0.33 (0.15)

Standard errors in brackets,
in first column from replications.

from the next value and therefore clearly out of line. There remain
20 189 good loans and 627 bad loans. A logit has been fitted to reduced
samples of equal numbers (with 627 good loans drawn at random) and
this gives quite different estimates from the full sample. As the result
may be due to chance, we have repeated the procedure 100 times; but
Table 6.2 shows there remains an appreciable difference between the two
estimates. In this case the one-sided sample reduction does not work.
This brings home that the short-cut method works only in the special
case of a logit probability. But the present data do not satisfy the logit
model, as we have seen in Section 4.3.

When the method does work the question arises *how many* zero observations should be included in the reduced sample. In applied work
there is a strong tradition of using equal numbers from both categories,
putting $l^{\bullet} = m$, presumably for reasons of symmetry. The precision of
the estimates will improve with a larger number of zero observations,
but with such a one-sided increase the usual rule that variances vary in-
versely with sample size does not apply. At this point the analogy with
discriminant analysis is helpful for a rough approximation. By (6.3) logit
slope coefficients reflect the difference between the two population means
of the covariate, and by analogy estimated coefficients reflect the differ-

ence between sample means. Writing σ_1^2 and σ_0^2 for the within-group variances of a covariate X_j, elementary sampling theory gives

$$\text{var } \hat{\beta}_j = \text{var } (\bar{X}_{j1} - \bar{X}_{j0}) = \frac{\sigma_1^2}{m} + \frac{\sigma_0^2}{l^\bullet}.$$

If the within-group variances are assumed equal this gives

$$\text{var } \hat{\beta}_j \propto \frac{1}{m} + \frac{1}{l^\bullet}.$$

We now vary the number of zero observations by multiples K of the given number m, or $l^\bullet = Km$. This gives

$$\text{var } \hat{\beta}_j \propto 1 + \frac{1}{K}.$$

Finally we take the ratio of the variance to its value in the reference case with $K = 1$ and obtain

$$\frac{\text{var } \hat{\beta}_{j,K}}{\text{var } \hat{\beta}_{j,1}} = \frac{1}{2} + \frac{1}{2K}.$$

It follows that one-sided increases in the sample size will not reduce the variance below one half of its value for equal numbers. Most of the gain in precision is moreover obtained in the initial stages: K equal to 5 already realizes 80% of this potential gain. The third column of Table 6.1 shows that for this value the standard deviations are indeed quite close to the values for the full sample, where K is almost 200. This is in line with the view of Breslow and Day (1980, p. 27) that there is little additional precision to be gained by going beyond values of K of 3 or 4.

6.3 State-dependent sampling

The arguments of the last section also apply if the full sample coincides with the population, as in the case of companies quoted on the stock exchange or the loans of a particular bank. It is then a small step to consider *endogenous stratification*, with sampling rates that vary according to the outcome, or *state-dependent* or *choice-based* samples, composed of separate samples drawn from population segments with different outcomes. If the attribute under consideration is rare this form of data collection is much more efficient than random sampling, for a random sample must be very large to yield sufficient observations with the attribute. We give an example in the next section. Moreover, there are gains from adapting the data collection methods to the conditions of each stratum, as in transport studies where bus passengers are questioned at bus terminals, car travellers at car parks, and so on. Differences in collection costs between these strata can also be taken into account in determining the optimal allocation of effort. For more on this (in a different context) see Gail et al. (1976).

The argument of the last section applies to logit analyses of such samples. With endogenous stratification the sampling fractions for the two outcome strata are λ_1 and λ_0, and (6.5) is replaced by

$$P_i = \frac{\lambda_1 P_i^*}{\lambda_1 P_i^* + \lambda_0(1 - P_i^*)}.$$

If P^* is a logit model (6.6) is adjusted accordingly with the ratio λ_0/λ_1 replacing λ; the preceding section deals with the special case $\lambda_1 = 1$, $\lambda_0 = \lambda$. For logit models the conclusion still holds that the probabilities in the combined sample obey a logit model with the same slope coefficients as in a random sample, and an adjusted intercept

$$\beta_0 = \beta_0^* - \log(\lambda_0/\lambda_1).$$

This is an irresistible argument for assuming a logit model and employing standard maximum likelihood techniques for its estimation from a state-dependent sample or from separate samples from several sources. But if the phenomenon under consideration does not satisfy the plain logit model the technique is not applicable, as was demonstrated by the bank loan example of the preceding section. The case has been forcefully argued by Xie and Manski (1989).

Various alternative methods of estimation have been put forward for the general case. Manski and Lerman (1977), who were the first to raise this issue, show that the standard ML estimates are inconsistent under state-dependent sampling. They propose a *weighted maximum likelihood* method of estimation. If s denotes a state or outcome with frequency $K(s)$ in the population and $H(s)$ in the combined sample, the composition of the state-dependent sample is distorted by factors $H(s)/K(s)$. This can be undone by (re)weighting the loglikelihood of the individual observations $\log Pr(Y_i)$ by the inverse $K(s)/H(s)$. The method is intuitively appealing: in the one-sided sample reduction case it would mean that we take a fraction λ of the abundant outcome, and then replicate these observations by a factor $1/\lambda$ in the calculations. But these weights are not always known: while $H(s)$ is obviously known, $K(s)$ is not. Approximate values may sometimes be obtained from outside sources, but the $K(s)$ may also have to be inferred from the sample data as part of the estimation procedure, which complicates matters considerably. For a review of the theory of this (and related) methods of estimation we refer to Amemiya (1985, Ch. 9.5), and for an appraisal of the performance of the various estimators in the case of logistic regression to Scott and Wild (1986).

Other methods of estimation are Manski's maximum score method, originally proposed as a nonparametric or distribution-free method of estimation, from Manski (1985), and the moment estimator of Imbens (1992). But this is beyond our scope.

6.4 Case–control studies

Case-control studies or *retrospective* analyses can be regarded as a special form of the state-dependent sampling methodology of the previous section. This approach has been developed independently and vigorously in epidemiology or medical statistics, with an agenda and a terminology of its own. We shall here present it within the framework of the logistic regression model of Chapter 2, but this will only cover the simplest form. Great advances have been made in extending the method to *matched samples*, and in the combination of evidence from several independent case–control studies. The reader is referred to the survey of Breslow (1996) as an entry to the vast literature in this field.

In case–control studies the distinction between *cases* and *controls* refers to the dichotomy by outcome: the cases are patients who suffer from a particular disease, and the controls are individuals who do not. In the basic form there is only a single discrete covariate, a categorical variable like *exposure* to adverse conditions. The data consist of two separate samples: the cases are as a rule at hand in a hospital, or in the files of a medical consultant, and the controls are taken from any data readily available. The natural tendency is to have approximately equal numbers of both categories, presumably for reasons of symmetry, although the merits of this usage can be challenged. The controls are usually selected with a view to similarity to the cases in respect of major characteristics like age, gender, and habitat; if the cases are patients with breast cancer, the controls will also be adult women. This mild control of the sample composition by two or three major factors is generally thought a sufficient precaution to justify the neglect of other covariates than the single exposure variable. While the data are treated as if they were random samples from the two populations, they have seldom been obtained in this manner.

Analyses of separate samples

Table 6.3. *A* 2×2 *table.*

	$Y = 0$	$Y = 1$
$X = 0$	n_{00}	n_{01}
$X = 1$	n_{10}	n_{11}

In the simple version we consider here with a single binary covariate the data can be summarized in a 2×2 contingency table by outcome and exposure, with $Y = 1$ for the cases and $Y = 0$ for the controls, and a categorical covariate $X - 1$ and $X = 0$ for the presence or absence of exposure. It makes no difference whether we write the observations as (relative) frequencies or as numbers, as we do in Table 6.3.

The central concept of case–control studies is the *odds*; the odds by exposure are n_{01}/n_{00} and n_{11}/n_{10} respectively, and the *odds ratio* is

$$\frac{n_{01}}{n_{00}} \frac{n_{10}}{n_{11}}. \tag{6.7}$$

This is a measure of the effect of exposure; if the incidence of the outcome is small, that is if the attribute or disease under consideration is rare in both exposure classes (as is often the case), both odds are close to zero and therefore a close approximation of the corresponding *risks*

$$\frac{n_{01}}{n_{00}} \approx \frac{n_{01}}{(n_{00} + n_{01})},$$
$$\frac{n_{11}}{n_{10}} \approx \frac{n_{11}}{(n_{10} + n_{11})}.$$

In this case the odds ratio is approximately equal to the relative risk in the exposed population.

It is remarkable that the same odds ratio holds if the roles of X and Y are reversed, and we start from the odds of exposure in the two outcome classes: if we take odds in the columns instead of the rows of Table 6.3, and proceed as above, the same odds ratio is obtained.

The effect of exposure is often expressed by taking the logarithm of the odds ratio of (6.7). This is equivalent to the slope coefficient in a standard binary logit model with a single categorical covariate, and we shall regard it in this light. In that model we have

$$\mathrm{logit}[P(Y = 1|X)] = \alpha + \beta X,$$

for the logit or log odds. If X takes only two values, 0 and 1, there are only two probabilities or frequencies, and only two log odds or logits, and these are

$$\log(n_{01}/n_{00}) = \alpha,$$
$$\log(n_{11}/n_{10}) = \alpha + \beta.$$

The *log odds ratio* is the difference of these logits, and corresponds to β; it provides an estimate of that coefficient which reads

$$\hat{\beta} = \log\left(\frac{n_{11}/n_{10}}{n_{01}/n_{00}}\right). \tag{6.8}$$

As we have seen in the preceding section, the use of a state-dependent or retrospective sample affects the intercept but not the slope coefficient β, and maximum likelihood estimation on the lines of Section 3.3 is therefore applicable. It gives the same result as (6.8). This is one of the rare cases that the maximum condition (3.20) does permit of an analytical solution. It gives rise to two equations in the elements of the 2×2 table and the probabilities $Pl(Y = 1)$ and $Pl(Y = 0)$, and when these are solved this gives the same $\hat{\beta}$ as above.

We have already pointed out that the same odds ratio (6.7) holds if the roles of X and Y are reversed. Hence the effect of Y on X is identical to the effect of X on Y: if we consider a logit model the other way round, say

$$\text{logit}\,[P(X = 1|Y)] = \alpha^o + \gamma Y,$$

the estimate $\hat{\gamma}$ is identical to $\hat{\beta}$ of (6.8), for we find

$$\hat{\gamma} = \log\left(\frac{n_{11}/n_{01}}{n_{10}/n_{00}}\right).$$

Since $\hat{\beta}$ from the log odds ratio is identical to the maximum likelihood estimate, its variance can also be established from that theory by implementing the information matrix $\mathbf{H}$ of (3.17) for the particular data structure of the present case, and then inverting it, as in (3.8). In the present instance $\mathbf{H}$ is a 2×2 matrix, and this exercise is not particularly arduous; it gives

$$\text{var}\,\hat{\beta} = \frac{1}{n_{00}} + \frac{1}{n_{01}} + \frac{1}{n_{10}} + \frac{1}{n_{11}}. \tag{6.9}$$

The one-sided sample reduction, endogenous stratification and state-dependent sampling of the last three sections have all been prompted by concern over the small number of rare attributes which will turn up in a

Analyses of separate samples

random sample. We conclude by an empirical illustration of the various sampling schemes with an application to 2×2 contingency tables of cases and controls. In this example the cases (with $Y = 1$) are motorists who have sustained severe injuries from a traffic accident, and the exposure variable X is the vehicle mode: $X = 1$ for riding a motorcycle versus $X = 0$ for travelling by car or truck (anything with four or more wheels), for passengers and drivers alike. The starting point is a known population, with aggregate data from the Netherlands in 1994, taken from the official statistics of Centraal Bureau voor de Statistiek (1997); this forms the top panel of Table 6.4. The following panels show the expected numbers in contingency tables from hypothetical samples of various types. Since they are *expected* numbers, all these tables give the same case–control

Table 6.4. *Motor traffic participants with severe injuries: expected numbers in various samples.*

	$Y = 0$	$Y = 1$
Population		
$X = 0$	8 820 828	5172
$X = 1$	320 213	1087
Random sample, size 8500		
$X = 0$	8197	5
$X = 1$	297	1
Stratified sample, twice 4250		
$X = 0$	4248	2
$X = 1$	4236	14
Retrospective sample, twice 4250		
$X = 0$	4101	3512
$X = 1$	149	738
Retrospective sample, 7000 and 1500		
$X = 0$	6755	1240
$X = 1$	245	260
Case-control sample, twice 50		
$X = 0$	48	41
$X = 1$	2	9

estimate of $\hat{\beta}$ of (6.8), apart from the effect of rounding the sample frequencies to the nearest integer. This estimate is 1.75, so that the risk of injury while riding a motorcycle is exp(1.75) $\approx$ 5.75 times as great as for other motorists. But while the estimate is the same, its standard deviation from (6.9) varies widely from one sampling design to another. Inspection shows that it is very sensitive to small numbers of observations in any one cell. What we want is therefore sizeable frequencies in all cells, or, if we wish to economize on the overall sample size, a balanced distribution of observations over the four cells.

The first case is a random sample of size 8500 from the entire population. This very large number is dictated by the need to have an expected number of at least one observation in each cell; there is of course still a definite risk that sampling will produce a zero cell which will wreck the analysis altogether (see Section 3.4). In spite of the large overall sample size, it is a very inefficient design, and the standard deviation of $\hat{\beta}$ is 1.1. The next panel describes a sample of the same size with exogenous stratification, composed of two samples of size 4250 from the two exposure categories, motorcyclists. This does a little better in respect of sparse cells, but not much; the standard deviation of $\hat{\beta}$ is 0.8.

The next two samples are retrospective or state-dependent samples of the same size, and they show a spectacular improvement. The first consists again of twice 4250 observations, but now from each of the outcome classes, and this reduces the standard deviation of $\hat{\beta}$ to 0.1. It might be thought that further improvement would be achieved by having unequal samples with a view to more balanced cell numbers, but in the present instance this makes no difference to the standard deviation of $\hat{\beta}$.

In the interest of comparability we have adopted the same overall sample size of 8500 throughout, but this is a very large number for any survey. If a genuine case–control study were undertaken by an emergency room coping with traffic accidents the number of cases and controls would probably be around 50. The expected numbers are given in the last panel of the table; they give a standard deviation of $\hat{\beta}$ of 0.85.

7

The standard multinomial logit model

This chapter and the next deal with extensions of the binary model to more than two outcomes. This chapter treats ordered probability models very briefly and the standard multinomial logit model at some length. The ordered probability models are a direct and fairly narrow generalization of the stimulus and response models of Section 2.2, and in particular of the latent regression equation of Section 2.3; the standard multinomial logit is a direct generalization of the binary logistic *formula*, without reference to any particular underlying idea. This model differs much more sharply from the binary model, and it is more versatile than the ordered model. Its properties as well as its estimation deserve a fuller explanation. We also give an empirical application, once more to household car ownership. The chapter is concluded by a a test for pooling states (and thereby reducing the number of distinct outcomes).

There are still other generalizations of the binary model to more than two alternatives, for this has been the chosen vehicle for more profound theories of choice behaviour which bring new practical implications with them. This is the subject of Chapter 8.

7.1 Ordered probability models

We recall the simple stimulus and response models and their formal representation in the latent regression equation model of Section 2.3. The stimulus determines a latent variable Y_i^* by the ordinary regression equation (2.8),

$$Y_i^* = \beta_0^* + \tilde{\mathbf{x}}_i^T \tilde{\boldsymbol{\beta}}^* + \varepsilon_i^*, \qquad (7.1)$$

104

where we now distinguish the intercept from the slope coefficients. The observed dichotomous outcome Y_i is determined by the sign of Y_i^*, as in (2.9):

$$Y_i = 1 \ \text{if } Y_i^* > 0,$$
$$Y_i = 0 \ \text{otherwise.}$$

This is equivalent to the operation of a zero threshold for the latent variable. A straightforward generalization is to allow for a graduated response, distinguishing several ordered states r which correspond to successive intervals of Y_i^*. Instead of recording survival and death of insects subjected to a toxin (or of patients admitted to an intensive care unit), we allow for intermediate categories of continuing illness or invalidation. Marketing firms will distinguish several sections of the public, from those who never buy their product to customers with an almost addictive loyalty. Ordered categories of this sort are particularly popular in marketing and public opinion surveys where respondents are invited to express their feelings on particular issues on a five-point or seven-point scale.

The R ordered categories correspond to intervals of Y_i^*, separated by thresholds α_r^*, and the outcome is denoted by a vector $\mathbf{y}_i$ which has zeros everywhere apart from a single 1 at the position indicating the state $r(i)$ that actually obtains. Formally we have

$$r(i) = 1 \ \text{if } -\infty < Y_i^* < \alpha_1^*,$$
$$r(i) = t \ \text{if } \alpha_{t-1}^* < Y_i^* < \alpha_t^*,$$
$$r(i) = R \ \text{if } \alpha_{R-1}^* < Y_i^* < \infty.$$

It is clear from (7.1) that the intercept β_0^* and the α_r^* are not separately identified. In the binary case this was resolved by setting the threshold at zero; here we put the intercept at zero and retain the thresholds. Substitution gives the inequality conditions in terms of the random disturbance ε_i^*, as in

$$r(i) = t \ \text{if } \alpha_{t-1}^* - \tilde{\mathbf{x}}_i^T \tilde{\boldsymbol{\beta}}^* < \varepsilon_i^* < \alpha_t^* - \tilde{\mathbf{x}}_i^T \tilde{\boldsymbol{\beta}}^*.$$

This gives the probability of the observed outcome as a difference between two values of the distribution function of ε^*, and once again a further normalization of β and of the α_r with respect to its standard deviation is in order, exactly as in Section 2.3. The upshot is that we have

$$\Pr[r(i) = t] = P_{it} = F(\alpha_t - \tilde{\mathbf{x}}_i^T \tilde{\boldsymbol{\beta}}^*) - F(\alpha_{t-1}^* - \tilde{\mathbf{x}}_i^T \tilde{\boldsymbol{\beta}}^*), \qquad (7.2)$$

where $F(\cdot)$ is a given distribution function of ε with zero mean and fixed variance. This may again be a standardized normal or logistic distribution, but in practice most ordered probability models are ordered probits, not logits.

The ordered probit was introduced by Aitchison and Silvey (1957) and later in the present form by McKelvey and Zavoina (1975). Its use is mainly confined to marketing and public opinion analyses. The estimated coefficients β_j reflect the effect of the covariates X_j on the latent variable Y_i^*, and one can trace through (7.2) how this will affect the probabilities of the ordered alternatives. But this does not lead to a simple overall measure of the effect of covariates on the outcome. Odds may be defined, for example as the ratio

$$\Pr[r(i) \leq t]/\Pr[r(i) > t],$$

and for a logit specification the log odds are again linear in $\mathbf{x}_i$, but it is questionable whether this is of much use. And it is certainly difficult to use this model (or any other multinomial model) to assign observations to a certain class or group on the basis of estimated probabilities. With R probabilities attached to each $\mathbf{x}_i$, it is difficult to define an analogue to the cut-off criterion of a binary model (see Section 2.4). If we assign an observation to the class with the highest probability, this is an even worse prescription than the conventional cut-off point of 0.5 in the binary case: for the highest probability may be only a little in excess of $1/R$.

With k covariates and R classes the ordered probit model has $k + R - 1$ parameters, and their maximum likelihood estimation is a fairly straightforward exercise; the formulae can be found in Maddala(1983, Ch. 2.13). Some authors worry about the possibility that the estimates of the α_r do not obey the necessary ordering, but in practice this does not seem to be a problem.

7.2 The standard multinomial logit model

In multinomial probability models, also known as polychotomous or polytomous models, there are any number S of alternative outcomes or states with index $s = 1, 2, \ldots, S$. These states are disjoint and exhaustive: they cover all possible outcomes, if necessary by the addition of a residual category. Instead of treating private car ownership as a simple attribute we can distinguish ownership of new cars, of used cars, and of more than one car, or we may classify cars by price, age, size or engine power. For travellers making a particular trip, the choice of a mode of transport (or *modal split*) is between walking, riding a bicycle, using a private car, and public transport. These classifications are dictated by the purpose of the analysis and often limited by the nature of the data. In principle, categories can be distinguished at will, but some distinctions may be irrelevant to the choices under consideration. Section 7.6 provides a statistical test for pooling states. In spite of what the car ownership example suggests, the states are not ordered from 'less' to 'more', and if there is such an ordering it is disregarded. The standard multinomial model treats all states on the same footing and is impervious to changes in their order.

We denote the outcome at observation i by $\mathbf{y}_i$, a vector of S elements y_{is} with a single element equal to 1 and all others 0. The position of this unit element indicates the state that obtains or $s(i)$. The model determines a vector $\mathbf{p}_i$ of S probabilities

$$P_{is} = \Pr(y_{is} = 1) = P_s(\mathbf{x}_i, \boldsymbol{\theta})$$

for each trial or observation i as a function of covariates $\mathbf{x}_i$ and unknown parameters $\boldsymbol{\theta}$. In some cases, the range of feasible or accessible alternatives (or *choice set*) varies from one individual or trial i to another; in an analysis of modal split, the choice of some individuals is constrained because they have no car or because there is no public transport for their itinerary.† In these cases there is a known choice set $\mathcal{S}_i$ associated with each i, the probabilities are only defined for $s \in \mathcal{S}_i$, and the notation changes accordingly. As this complication seldom arises in connection with the standard logit model we shall ignore it in this chapter.

In the *standard multinomial logit model* the probability function is

$$P_{is} = P_s(\mathbf{x}_i, \boldsymbol{\theta}^*) = \frac{\exp(\mathbf{x}^T \boldsymbol{\beta}_s^*)}{\sum_{t=1}^{S} \exp(\mathbf{x}^T \boldsymbol{\beta}_t^*)}, \qquad (7.3)$$

with separate parameter vectors $\boldsymbol{\beta}_s^*$ for each state s. The $\mathbf{x}_i$ of $k + 1$ covariates always include a unit element, and the $\boldsymbol{\beta}_s^*$ an intercept. This model is clearly overparametrized, for we may add a constant vector to each $\boldsymbol{\beta}^*$ without affecting the P_{is}; the parameters are only determined up to an additive constant, they are not identified. But their differences $\boldsymbol{\beta}_s^* - \boldsymbol{\beta}_t^*$ are identified, and the standard remedy is to suppress one vector

---

† In extreme cases, the set is reduced to a single option, the individual has no choice, and probability models do not apply.

β_1^* by subtracting it from all β_s^*, reducing β_1^* itself to $\mathbf{0}$. Since the states may be (re)ordered at will it does not matter what state is taken as this *reference state* $s = 1$. The identifiable parameters are redefined as

$$\beta_s = \beta_s^* - \beta_1^*,$$
$$\beta_1 = \mathbf{0}. \tag{7.4}$$

and the probabilities as

$$P_{is} = P_s(\mathbf{x}_i, \boldsymbol{\theta})$$
$$= \frac{\exp(\mathbf{x}^T \beta_s)}{1 + \sum_{t=2}^{S} \exp(\mathbf{x}^T \beta_t)} \text{ for } s \neq 1,$$
$$P_{i1} = \frac{1}{1 + \sum_{t=2}^{S} \exp(\mathbf{x}^T \beta_t)}. \tag{7.5}$$

With S states and $k + 1$ elements in β_s the total number of parameters of the entire model is $(k + 1) \times (S - 1)$. Equations (7.3) and (7.5) are equivalent representations of the same probabilities, denoted in the present chapter by the shorthand notation Pl_s^* and Pl_s. (7.3) with its nice symmetry in the S states is best suited for theoretical discussions, but the reduced form (7.5) is in order when it comes to estimation and practical implementation.

For $S = 2$ (7.5) reduces at once to the binary model of (2.1), with $Y = 0$ the reference state. The present model is thus a straightforward algebraic generalization of the binary logit, and it was first presented in this manner by Cox (1966) and Theil (1969) (who incidentally also introduced elements of the conditional logit of Section 8.3). But as we have seen in the last section, a generalization of the underlying arguments of stimulus and response and of the latent regression leads to the ordered probability model, not to the present model. The nearest multinomial analogue to the latent regression model is probably the *discrete choice model*, in which the sophisticated and elegant formulations of random utility maximization replace the crude utility maximization of Section 2.2. This theory can very well serve to justify the standard multinomial, but in practice it is inextricably bound up with the conditional model. Its discussion is postponed to Chapter 8.

There is of course also a multinomial probit model, which initially ran in tandem with the multinomial logit. It is more flexible, as it allows naturally for correlation among the random elements, but analytically it is much less tractable and this has limited its practical use for some time. But new methods (and new computer capacities) have overcome this obstacle. We return to this subject in Section 8.5.

Proper odds make little sense in a multinomial context: their analogue is the ratio of two probabilities of any pair of states (s, t). Its logarithm closely resembles the log odds or logit of (2.4)

$$R(s, t) = \log(P_s/P_t)$$

which gives

$$R(s, t) = \mathbf{x}^T(\boldsymbol{\beta}_s - \boldsymbol{\beta}_t), \tag{7.6}$$

or, with t the reference state 1,

$$R(s, 1) = \mathbf{x}^T\boldsymbol{\beta}_s. \tag{7.7}$$

The linearity of all these pairwise log odds in the covariates $\mathbf{x}$ is a distinguishing characteristic of the present model, just as the linearity of the logit defines the binary model; when this property is taken as a starting-point, (7.5) will follow. Note that the log odds depends exclusively on the parameters of the two states concerned, regardless of all others. This property is known as the *independence from irrelevant alternatives*. It applies to the wider class of the general logit model, and it has major consequences for the behavioural implications of the discrete choice theory. For a fuller discussion see Section 8.1.

The standard multinomial probability does *not* share the monotonic behaviour of the binary probability. The derivative of Pl_s with respect to the jth regressor is

$$\partial Pl_s^*/\partial X_j = Pl_s^* \left(\beta_{sj}^* - \sum_{t=1}^{S} \beta_{tj}^* Pl_t^* \right),$$

or equally

$$\partial Pl_s/\partial X_j = Pl_s \left(\beta_{sj} - \sum_{t=1}^{S} \beta_{tj} Pl_t \right). \tag{7.8}$$

Inspection shows that the derivative is not affected by the choice of the reference state. By the values of the Pl_s and Pl_t, its value depends on the point of evaluation, just as in the bivariate case; but now it can also vary in sign between one point of evaluation and another, for the sign of the second term may change through changes in the probabilities. The multinomial logit probabilities may thus exhibit non-monotonic behaviour with respect to the elements of $\mathbf{x}$: we shall see in Section 7.5 that the probability of owning a used car first increases and then declines as income rises. This is an exceptional case, and as a rule probabilities will not change sign within the sample range of the regressors. Even so, it should be noted that the sign of the derivative (7.8) is not determined by $\beta_{sj} = (\beta_{sj}^* - \beta_{lj}^*)$ alone. The sign and size of covariate effects on the various probabilities cannot be inferred from the β_{sj}.

As before, the derivatives can be turned into quasi-elasticities

$$\eta_{sj} = X_j \partial Pl_s / \partial X_j,$$

which indicate the percentage point change in Pl_s upon a 1% increase in X_j. Over all states, the probabilities sum to 1, and the derivatives and quasi-elasticities to 0. Like the derivatives, quasi-elasticities are invariant to the choice of the reference state, and they may change in sign and size when they are evaluated at different points.

7.3 ML estimation of multinomial models: generalities

As in Section 3.1, we apply the principles of maximum likelihood estimation to a general multinomial probability model, not necessarily of the logit type; in the next section these formulae are adapted to the special case of the standard multinomial. Here, we start off from the S probabilities

$$P_{is} = P_{is}\left(\mathbf{x}_i, \boldsymbol{\theta}\right)$$

with $\mathbf{x}_i$ a vector of covariates and $\boldsymbol{\theta}$ a vector of K parameters. These probabilities are nonnegative and sum over i to 1 for all $\mathbf{x}$ and all $\boldsymbol{\theta}$, so that in particular

$$\sum_s P_{is} = 1 \text{ for all } i.$$

Moreover

$$\mathrm{E}y_{is} = P_{is}.$$

Assume that the maximum likelihood estimates are again calculated by the iterative scoring method of (3.10) of Section 3.1,

$$\boldsymbol{\theta}_{t+1} = \boldsymbol{\theta}_t + \mathbf{H}(\boldsymbol{\theta}_t)^{-1}\mathbf{q}(\boldsymbol{\theta}_t). \tag{7.9}$$

In addition to the loglikelihood function itself we shall need the score vector $\mathbf{q}$ of its first derivatives as well as the information matrix $\mathbf{H}$. We consider these elements in turn.

As always we assume independence of the observations, so that the sample loglikelihood is simply the sum over i of the loglikelihoods of the single observations; so are its first and second derivatives. The loglikelihood for observation i is

$$\log L_i = \sum_s y_{is} \log P_{is}.$$

As there is only a single nonzero y_{is}, this summation actually yields only a single term; the same holds for the expressions for q_{ij} and Q_{ijh} that follow. The sample loglikelihood is then

$$\log L = \sum_i \log L_i = \sum_i \sum_s y_{is} \log P_{is}.$$

A typical element q_j of the score vector $\mathbf{q}$ is the derivative of $\log L$ with respect to the jth element of $\boldsymbol{\theta}$. For observation i this is

$$q_{ij} = \sum_s \frac{y_{is}}{P_{is}} \frac{\partial P_{is}}{\partial \theta_j},$$

or, for the entire sample.

$$q_j = \sum_i \sum_s \frac{y_{is}}{P_{is}} \frac{\partial P_{is}}{\partial \theta_j}. \tag{7.10}$$

For the contribution of observation i to a typical element of the Hessian $\mathbf{Q}$, consider the second derivative of the loglikelihood function

$$Q_{ijh} = \sum_s \left(\frac{y_{is}}{P_{is}} \frac{\partial^2 P_{is}}{\partial \theta_j \partial \theta_h} - \frac{y_{is}}{P_{is}^2} \frac{\partial P_{is}}{\partial \theta_j} \frac{\partial P_{is}}{\partial \theta_h} \right),$$

which again yields only a single term. For the information matrix we reverse the sign and take the expected value by substituting $Ey_{is} = P_{is}$, and now the summation is no longer trivial. The sum of the first term turns into a sum of second derivatives, which vanishes since the P_{is} sum to 1 and the first derivatives sum to 0. The second term is much simplified: this yields

$$H_{ijh} = \sum_s \frac{1}{P_{is}} \frac{\partial P_{is}}{\partial \theta_j} \frac{\partial P_{is}}{\partial \theta_h}. \tag{7.11}$$

The value for the entire sample is of course

$$H_{jh} = \sum_i \sum_s \frac{1}{P_{is}} \frac{\partial P_{is}}{\partial \theta_j} \frac{\partial P_{is}}{\partial \theta_h}. \tag{7.12}$$

These expressions can be rewritten in matrix notation. The derivatives of the P_{is} with respect to the elements of $\boldsymbol{\theta}$ can be arranged in

$S \times K$ matrices $\mathbf{A}_i$, and the S probabilities P_{is} in diagonal $S \times S$ matrices $\check{\mathbf{p}}_i$. The rank of the $\mathbf{A}_i$ is at most $S - 1$, for as the P_{is} add up to 1 their derivatives add up to 0, and so do the columns of $\mathbf{A}_i$. We shall indeed assume that the rank of $\mathbf{A}_i$ *is $S - 1$* as we may reasonably suppose that the number of parameters K is at least equal to $S - 1$; in actual practice it is usually much larger. The information matrix for a single observation of (7.11) can then be rewritten as

$$\mathbf{H}_i = \mathbf{A}_i^T \check{\mathbf{p}}_i^{-1} \mathbf{A}_i.$$

This $K \times K$ matrix is singular, for its rank is set by the rank of A_i, and this is only $S - 1$.

The $\mathbf{H}_i$ are summed to the sample information matrix of (7.12)

$$\mathbf{H} = \sum_i \mathbf{A}_i^T \check{\mathbf{p}}_i^{-1} \mathbf{A}_i$$

and this operation can again be expressed in matrix terms. We stack the A_i in an $(n\,S) \times K$ matrix $\mathbf{A}$ as in

$$\mathbf{A} = \begin{pmatrix} \mathbf{A}_1 \\ \mathbf{A}_2 \\ \cdot \\ \cdot \\ \cdot \\ \mathbf{A}_n \end{pmatrix}$$

and the $\check{\mathbf{p}}_i$ are arranged in an $S \times n$ block diagonal matrix $\check{\mathbf{p}}$

$$\check{\mathbf{p}} = \begin{pmatrix} \check{\mathbf{p}}_1 & 0 & \dots & 0 \\ 0 & \check{\mathbf{p}}_2 & \dots & 0 \\ & & \dots & \\ & & \dots & \\ & & \dots & \\ 0 & 0 & \dots & \check{\mathbf{p}}_n \end{pmatrix}.$$

The end result is

$$\mathbf{H} = \mathbf{A}^T \mathbf{p}^{-1} \mathbf{A}. \tag{7.13}$$

$\mathbf{H}$ is a $K \times K$ matrix, and its rank is set by the ranks of its constituent parts. $\mathbf{p}$ has full rank, just like the $\mathbf{p}_i$; as for $\mathbf{A}$, its rank is determined by its number of columns K. Linear dependence among these columns is ruled out, for it would mean underidentification. The number of rows is of course far in excess of K: there are $n(S - 1)$ rows, and it is most unlikely that the rank is reduced below K by linear dependencies between the derivatives for different observations i.

If $\mathbf{A}^T$ is premultiplied by a nondegenerate $1 \times K$ row vector $\mathbf{z}^T$, this will give a column vector of length nS; pre- and postmultiplication of $\mathbf{H}$ by $\mathbf{z}$ therefore yields a weighted sum of squares with positive weights $1/P_{is}$, which are positive scalars. This means that $\mathbf{H}$ is a positive semidefinite matrix for any feasible set of parameter values that produces nonnegative probabilities, always provided K exceeds $S - 1$; we must have at least one parameter for each probability, bearing in mind that these sum to 1. But if $\mathbf{H}$ is positive definite, the scoring algorithm (7.9) will always converge to a maximum, whatever the starting values. The argument moreover suggests (but does not prove) that the Hessian $\mathbf{Q}$ is negative definite for all parameter vectors, so that the loglikelihood function is everywhere convex and has a unique maximum. The implication is that we can trust the scoring algorithm to turn up with the proper ML estimates of the parameters. We assert these powerful properties without a proper proof; see McFadden (1974, p. 119) for the conditions for the Hessian of the logit model to be everywhere negative definite. – Note that the present convergence argument is quite general. It holds for any multinomial probability model, and therefore also for any binary model; for all varieties of the logit model, for probit models, and for any other probability model that may be devised.

7.4 Estimation of the standard multinomial logit

We now apply the general principles of the last section to the standard model (7.5),

$$Pl_{is} = Pl_s(\mathbf{x}_i, \boldsymbol{\beta})$$

$$= \frac{\exp(\mathbf{x}^T \boldsymbol{\beta}_s)}{1 + \sum_{t=2}^{S} \exp(\mathbf{x}^T \boldsymbol{\beta}_t)} \text{ for } s \neq 1,$$

$$Pl_{i1} = \frac{1}{1 + \sum_{t=2}^{S} \exp(\mathbf{x}^T \boldsymbol{\beta}_t)}.$$

With S states and $k + 1$ covariates (including a unit constant) the parameter vector $\boldsymbol{\beta}$ consists of $S - 1$ subvectors $\boldsymbol{\beta}_s$ of length $k + 1$ for $s = 2, 3, \ldots, S$.

No problems arise in the substitution of these probabilities into the loglikelihood function. The derivatives of $\mathbf{q}_i$ of (7.10), however, require some further algebra. To begin with, we must distinguish between the derivatives of Pl_{is} with respect to the elements of its 'own' parameter vector $\boldsymbol{\beta}_s$ and the derivatives with respect to elements of alien subvectors $\boldsymbol{\beta}_t$ with $t \neq s$. We find

$$\partial Pl_{is}/\partial \beta_j = X_{ji} Pl_{is}(1 - Pl_{is}) \text{ if } \beta_j \in \boldsymbol{\beta}_s,$$

$$\partial Pl_{is}/\partial \beta_j = -X_{ji} Pl_{is} Pl_{it} \quad \text{if } \beta_j \in \boldsymbol{\beta}_t, t \neq s.$$
(7.14)

For the reference state all parameter vectors are alien, and the second line applies with $s = 1$ to Pl_{i1}. We substitute these expressions into (7.10), with $\boldsymbol{\beta}_r$ denoting the subvector to which β_j belongs. This gives

$$q_{ij} = \frac{y_{is}}{P_{is}} X_{ji} Pl_{ir}(1 - Pl_{ir}) - \sum_{s \neq r} \frac{y_{is}}{P_{is}} X_{ji} Pl_{is} Pl_{ir}$$

$$= X_{ji}(y_{ir} - Pl_{ir} y_{ir} - Pl_{ir} \sum_{s \neq r} y_{is})$$

$$= X_{ji} (y_{ir} - Pl_{ir}),$$

where we make use of the fact that by their definition the y_{is} sum to 1. For the entire rth segment of $\mathbf{q}_i$, corresponding to $\boldsymbol{\beta}_r$, we have

$$\mathbf{q}_{ir} = (y_{ir} - Pl_{ir}) \mathbf{x}_i,$$

or, for the whole sample,

$$\mathbf{q}_r = \sum_i (y_{ir} - Pl_{ir}) \mathbf{x}_i.$$
(7.15)

Note that in the summation over i many y_{ir} will be zero: y_{ir} is only 1 for observations with state r. – These subvectors of the score vector have length $k + 1$ (since $\mathbf{x}_i$ has length $k + 1$), as they should, and there are $S - 1$ of them for $r = 2, 3, \ldots, S$. Stacked on top of one another they form the column vector $\mathbf{q}$.

It is seen from (7.15) that the random variables y_{ir} disappear upon further differentiation of $\mathbf{q}$ with respect to elements of $\boldsymbol{\beta}$. The Hessian matrix of second derivatives $\mathbf{Q}$ is therefore nonrandom, and for the information matrix $\mathbf{H} = -E\mathbf{Q}$ we need only reverse its sign. While $\mathbf{H}$ is easily obtained in this manner, we shall here find it by substituting the derivatives of (7.14) into (7.12), which gives a typical (j, h)th element of $\mathbf{H}$. As before, we must distinguish between the case that β_j and β_h both belong to the same subvector, say $\boldsymbol{\beta}_r$, and the case that they belong to different subvectors. $\mathbf{H}$ is a square matrix of order $K = (S-1)(k+1)$, and it can be partitioned into $(S-1)(S-1)$ submatrices of order $(k+1) \times (k+1)$, with the $\mathbf{H}_{r,r}$ on the main diagonal and the $\mathbf{H}_{r,t}$ the off-diagonal blocks.

First take the case that both β_j and β_h belong to the same subvector β_r. Here substitution of (7.14) into (7.11) gives

$$H_{ijh} = \frac{1}{P_{ir}} X_{ji} X_{hi} Pl_{ir} (1 - Pl_{ir})^2 + \sum_{s \neq r} \frac{1}{P_{is}} X_{ji} X_{hi} (Pl_{is} Pl_{ir})^2$$

$$= X_{ji} X_{hi} \left[Pl_{ir} (1 - Pl_{ir})^2 + \sum_{s \neq r} Pl_{is} Pl_{ir}^2 \right]$$

$$= X_{ji} X_{hi} Pl_{ir} (1 - Pl_{ir}). \tag{7.16}$$

In passing from the second to the third line we make use of the fact that the Pl_{is} sum over s to 1. If β_j and β_h belong to different subvectors a similar but slightly lengthier development gives

$$H_{ijh} = -X_{ji} X_{hi} Pl_{ir} Pl_{it}. \tag{7.17}$$

The submatrices on the diagonal are constructed from (7.16), and the others from (7.17); we find

$$\begin{aligned}
\mathbf{H}_{(rr)i} &= Pl_{ir}(1 - Pl_{ir})\mathbf{x}_i\mathbf{x}_i^T, \\
\mathbf{H}_{(rt)i} &= -Pl_{ir} Pl_{it}\mathbf{x}_i\mathbf{x}_i^T.
\end{aligned} \tag{7.18}$$

The complete matrix $\mathbf{H}$ is obtained by summing these expressions over i and arranging the blocks as indicated.

All the elements needed for estimation by the scoring algorithm (7.9) have now been assembled. For given parameter values β^0 the probabilities P_{is} are easily obtained, and these can be inserted in the formulae for the current values of $\mathbf{q}$ and $\mathbf{H}$ given above, and this is all we need. After convergence we obtain the same results as in the bivariate case, viz. parameter estimates, their asymptotic variances, and the value of $\log L$ at its maximum. Other statistics like derivatives and (quasi-)elasticities, with their standard errors, can be derived as before. The properties of maximum likelihood estimation listed in Section 3.1 apply as a matter of course.

In practice estimation will be carried out by means of some program package. The above algebra may help to understand what is going on, and occasionally what is going wrong. Much of the discussion of the binary logit estimation at the end of Section 3.3 applies equally here. Upon summing the expressions of (7.18), we again find that $\mathbf{H}$ closely resembles $\mathbf{X}^T\mathbf{X}$, the regressor moment matrix of ordinary regression; the difference lies in a fairly complicated weighting scheme, with various terms of the form $Pl_{ir} Pl_{it}$ as the weights. Still the main arguments about the structure of the regressor matrix $\mathbf{X}$ once more apply. It must have full rank and preferably little collinearity, and in order to facilitate its numerical inversion it should be well balanced in the sense that the diagonal elements of $\mathbf{X}^T\mathbf{X}$ are of the same order of magnitude. This is achieved by scaling the covariates so that they have comparable variances.

The MLEs of the parameters define ML predictions of the probabilities,

$$\widehat{Pl}_{is} = Pl(\mathbf{x}_i, \hat{\boldsymbol{\beta}}),$$

and these must satisfy the first-order conditions for a maximum $\mathbf{q}_s = \mathbf{0}$, or, by (7.15),

$$\sum_i (y_{is} - \widehat{Pl}_{is})\mathbf{x}_i = \mathbf{0}.$$

The first term on the left is again termed a *quasi-residual* e_{si}

$$e_{is} = y_{is} - \widehat{Pl}_{is},$$

or, arranging all three terms in matrices with n rows and S columns,

$$\mathbf{E} = \mathbf{Y} - \widehat{\mathbf{Pl}}.$$

In this notation, the first-order condition reads as

$$\mathbf{X}^T\mathbf{E} = \mathbf{0}. \tag{7.19}$$

Each row of $\mathbf{Y}$ consists of $S - 1$ zeros and one unit element, each row of $\widehat{\mathbf{Pl}}$ consists of nonnegative elements that sum to 1, and each row of $\mathbf{E}$ therefore sums to 0.† The columns of $\mathbf{Y}$ sum to the sample frequencies of the outcome states n_s, and by (7.19) the columns of $\mathbf{E}$ sum to 0 since $\mathbf{X}$ contains a column of unit constants. Hence

$$\sum_i e_{is} = 0,$$

and

$$\sum_i \widehat{Pl}_{is}/n = \sum_i y_{is}/n,$$

so that the *equality of the means* also holds in the multinomial case: the mean predicted probability of a state equals its sample frequency.

This also holds for the null or base-line estimate, when we reduce the arguments $\mathbf{x}_i^T\boldsymbol{\beta}_s$ to a vector of constants. The fitted probabilities $\widehat{Pl}_s^{\circ}$

† If we had consistently treated the two outcomes of the binary case on an equal footing, their quasi-residuals would be equal and of opposite sign.

are constants, too, and by the equality of the means they must be equal to

$$\widehat{Pl_s^\circ} = n_s/n.$$

The corresponding null loglikelihood is therefore

$$\log L^\circ = \sum_s n_s \log n_s - n \log n. \tag{7.20}$$

This can be used in an overall LR test of the performance of the regressor variables (other than the intercept) along the lines of (3.24).

Further analysis will bring to light that the *zero cell* complication of Section 3.4 may also occur in a multinomial context.

If there is only a limited number of regressor variables that are all measured in intervals or classes, the sample data consist of a cross-classification of all observations by these variables, with the frequencies of the S states in each cell. This is the case of categorical covariates. As in Section 3.5, the estimation presents no problems when the grouped observations are treated as repeated individual observations. It would be tedious to repeat the argument in detail. Minimum chi-squared estimation is also feasible, although I know of no examples. As for the traditional logit transformation of observed cell frequencies at the end of Section 3.5, its analogue in the multinomial case is the log of the ratio of any two state frequencies in the covariate cells. Let f_{js} denote the relative frequency of state s in cell j of some (cross-)classification by covariates, and f_{j1} the frequency of the reference state in that cell. If these frequencies reflect standard multinomial logit probabilities, the logarithm of their ratio is

$$\log(f_{js}/f_{j1}) \approx \mathbf{x}_j^T \boldsymbol{\beta}_s,$$

as in (7.7). This once more suggests ordinary regression of this log ratio. In economics such analyses are fairly common for *shares* of aggregates like expenditure or trade flows, even though these are not frequencies and have little or nothing to do with probabilities, apart from the fact that they are nonnegative and sum to 1.

7.5 Multinomial analysis of private car ownership

The standard multinomial logit model is illustrated by a continued analysis of private car ownership, based on the same data set as in Sections 3.6 and 3.7 and making use of the same regressors. We now distinguish three forms of private car ownership, or four states in all, viz.

Table 7.1. *Multinomial analysis of car ownership: effect of adding regressor variables on loglikelihood.*

nr.	regressors	$\log L$
0	constant only	-3528.37
1	constant, LINC	-3466.80
2	constant, LINC, LSIZE	-3176.80
3	constant, LINC, LSIZE, BUSCAR	-2949.76
4	constant, LINC, LSIZE, BUSCAR, AGE	-2884.59
5	constant, LINC, LSIZE, BUSCAR, AGE, URBA	-2874.90

- NONE, household does not own private car (1010 households);
- USED, household owns one used private car (944 households);
- NEW, household owns one new private car (691 households);
- MORE, household owns more than one private car (175 households).

NONE is taken as the reference state, but all results reported below are invariant to this choice.

Table 7.1 shows the course of the maximum loglikelihood as additional regressors are successively introduced, starting from the null model of a constant only. This shows the same unsteady rise as in Table 3.4 of Section 3.7, with large leaps forward upon the introduction of LSIZE and BUSCAR and quite small further contributions of AGE and URBA. We can test for the significance of additional regressors by the likelihood ratio test of (3.11); since there are four states and one is the reference state, there are three parameters associated with each covariate. The 5% significant value of chi-squared with three degrees of freedom is 7.815, so that the loglikelihood should increase by at least 3.9 for each additional regressor, as it amply does. – While the rise of the loglikelihood is similar to that in the binary case, its level is very much lower. This is due to the increased number of states. Recall that by (3.2) of Section 3.1 the maximum loglikelihood can be written as

$$\log L = \sum_i \log \widehat{\Pr}(Y_{is(i)})$$

with $\widehat{\Pr}(Y_{is(i)})$ the predicted probability of the observed state $s(i)$. Unless there is perfect discrimination, the probabilities are bound to become smaller as more states are distinguished, as the probability of a particular form of car ownership must be less than the probability of car ownership as such. The strength of this effect can be demonstrated for the loglikelihood of the base-line model of (7.20),

$$\log L^\circ = \sum_s n_s \log n_s - n \log n.$$

Consider a given set of S states, and let one state t be further subdivided into J states t_j; then $\log L^\circ$ changes by

$$\sum_j n_{tj} \log n_{tj} - n_t \log n_t$$

which is always negative. For the present subdivision of car ownership as such into three separate categories this amounts to -1688.75, as can be verified from the sample numbers given above.

Table 7.2 shows the effect of additional regressors on the derivatives with respect to LINC, or quasi-elasticities with respect to income per head. Note that the effect of income on the proportion of used car owners is almost negligible throughout. As more regressor variables are added, the t-ratios improve (as was to be expected), but the systematic movement away from zero of the binary model is not repeated; it appears that the omitted variables bias of Section 5.3 does not carry over to the multinomial case.

Table 7.2. *Multinomial analyses of car ownership:*
income effects.

nr.	NONE	USED	NEW	MORE
1	−0.08 (4.11)	−0.13 (6.36)	0.18 (10.00)	0.03 (2.61)
2	−0.43 (14.60)	−0.01 (0.47)	0.28 (11.44)	0.16 (11.35)
3	−0.59 (17.04)	0.06 (1.99)	0.34 (13.10)	0.19 (12.49)
4	−0.56 (16.28)	0.04 (1.31)	0.34 (12.87)	0.18 (12.02)
5	−0.56 (16.29)	0.04 (1.37)	0.34 (12.90)	0.18 (12.01)

Derivatives with respect to LINC,
absolute values of $t-ratios$ *in brackets.*

Table 7.3. *Multinomial analysis of car ownership:*
regressor effects on four states.

	NONE	USED	NEW	MORE
LINC	−0.56	0.04	0.34	0.18
	(16.29)	(1.37)	(12.90)	(12.01)
LSIZE	−0.67	0.23	0.20	0.25
	(19.83)	(7.56)	(7.60)	(13.98)
BUSCAR	−	−3.00	−3.02	−3.60
		(16.10)	(14.39)	(9.74)
URBA	0.03	−0.02	−0.01	−0.00
	(4.21)	(3.11)	(1.82)	(0.08)
AGE	0.03	−0.04	0.01	0.00
	(8.22)	(10.74)	(2.05)	(1.16)

Derivatives at the sample mean, except for
BUSCAR; *absolute values of t − ratio in brackets.*

The results for the full model with all five regressors are shown in
Table 7.3. For most covariates this gives the derivatives of (7.8); their
variances have been calculated according to (3.9). Since LINC and LSIZE
are in logarithms, their derivatives are quasi-elasticities. For BUSCAR
however we prefer the log probability ratio of (7.7) with respect to NONE.
Since BUSCAR is a $(0,1)$ dummy and NONE is the reference state, the
log probability ratios are given by the β_{3s}.† For all three types of car
ownership, the log probability ratio with respect to NONE is about -3, so
that the ratio of the probabilities is of the order of $\exp(-3) = 0.05$. All
car ownership probabilities are thus severely reduced by the presence
of a business car, in other words a business car is an almost perfect
substitute for any category of private car. – Income rises lead to a shift
towards new cars and multiple car ownership, but hardly affect used
car ownership; in contrast, increases in family size do cause a shift from
NONE to USED. URBA and AGE have only modest effects on all classes of
private car ownership.

All effects have been measured at the mean sample frequencies of the
four ownership states, and this gives a narrow view; when the regressors
range over wider intervals, the outcome can be quite different. In the
present case the low income elasticity of used car ownership arises largely

† With a different reference state, we would subtract the coefficients of NONE from
the β_{3s} to obtain the log probability ratio; for the t-ratio we would however need
the adjusted variances, and these must be constructed with the help of (3.9).

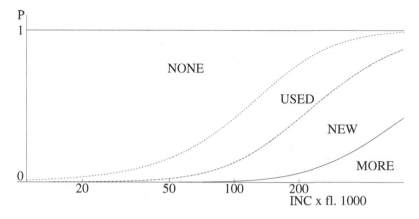

Fig. 7.1. Car ownership status as a function of household income.

because at the sample mean the probability of used car ownership is almost at its peak. Figure 7.1 shows how the shares of the four states shift
when income varies over a wide range, with all other regressor variables
kept constant at their sample mean values. The sample mean of INC is
about fl. 16000, or about $8000 (in 1980), and at this value USED is near
its maximum. The incomes at the lower end of the graph are unrealistic,
but they do illustrate how USED first increases with rising income at the
expense of NONE and then declines because it is overtaken by NEW. As
we have seen in connection with (7.8), multinomial probabilities, unlike
binary probabilities, do not vary monotonically with the covariates.

Another way of illustrating the estimates is to work out predicted
ownership rates for a few stereotype households. In Table 7.4, the first
column gives these probabilities at the sample mean covariates; they
differ a little from the sample frequencies, especially for MORE, because
of the nonlinearity of the probability functions. The other two columns
refer to two specimen households. Household A is a large family with a
young father, a modest income, living in the country; B is a much older
pair, quite well off and living in a large city. Neither household has a
business car. The contrast in car ownership probabilities is striking. The
poor countryside family is more likely to own a car than the rich city
dwellers, but it goes for used cars while the latter usually have a new
car. The differences between the two families may be decomposed and
attributed to the various aspects in which they differ. The exercise could
also be extended to a group of various households in given proportions,

Table 7.4. *Estimated car ownership probabilities*
at selected regressor values.

	sample mean	household A	household B
regressors			
income per head	15.74	6.0	25.0
size	2.12	4.0	1.7
business car	0.12	0	0
urbanization	3.74	1	6
age of head	45	22.5	52.5
ownership status			
NONE	33.6	14.3	24.5
USED	36.7	74.7	26.0
NEW	27.1	9.4	43.4
MORE	3.1	1.6	6.1

Geometric means; income per head in fl.1000 p.a.,
size in number of equivalent adults.

for example in order to assess the likely demand for parking places for an apartment building with a mixed group of inhabitants, and it would then amount to micro-simulation.

7.6 A test for pooling states

So far the classification of outcomes by states has been taken as given. As the analyst can distinguish as many states as the data permit, an understandable wish to preserve information can easily lead to irrelevant distinctions. In the interest of parsimony states should however be pooled together unless they are significantly different for the purpose of the analysis. This issue was first tackled by Hill (1983) in a study of female labour participation in underdeveloped countries, and a statistical test has been provided by Cramer and Ridder (1991, 1992).

We reason in reverse from the introduction of a superfluous distinction in an initial model with S states. State t (not the reference state) is singled out for further subdivision. In the light of the things to come we rewrite the standard formula (7.5) of Section 7.2 in a slightly unusual way, separating off state t and distinguishing intercepts β_{0s} and slope coefficients $\tilde{\boldsymbol{\beta}}_s$. This gives

$$P_{it} = \frac{\exp(\beta_{0t} + \tilde{\mathbf{x}}_i^T \tilde{\boldsymbol{\beta}}_t)}{1 + \exp(\beta_{0t} + \tilde{\mathbf{x}}_i^T \tilde{\boldsymbol{\beta}}_t) + R_i},$$ (7.21)

$$R_i = \sum_{s \neq 1, t} \exp(\beta_{0s} + \tilde{\mathbf{x}}_i^T \tilde{\boldsymbol{\beta}}_s).$$

Now suppose all outcomes of state t are further divided into two subsets by an arbitrary criterion like the name of a street or the colour of a bus (a classic example that will be used later in an entirely different context), the outcomes being allotted at random to the two new classes u and v in proportions λ and $(1 - \lambda)$. The original model is then extended to $S + 1$ states, and in lieu of P_{it} we have two probabilities

$$P_{iu} = \lambda P_{it}, \quad P_{iv} = (1 - \lambda) P_{it}.$$ (7.22)

All other probabilities remain the same. Substitution of (7.21) gives

$$P_{iu} = \frac{\exp(\log \lambda + \beta_{0t} + \tilde{\mathbf{x}}_i^T \tilde{\boldsymbol{\beta}}_t)}{1 + \exp(\beta_{0t} + \tilde{\mathbf{x}}_i^T \tilde{\boldsymbol{\beta}}_t) + R_i},$$

$$P_{iv} = \frac{\exp[\log(1 - \lambda) + \beta_{0t} + \tilde{\mathbf{x}}_i^T \tilde{\boldsymbol{\beta}}_t]}{1 + \exp(\beta_{0t} + \tilde{\mathbf{x}}_i^T \tilde{\boldsymbol{\beta}}_t) + R_i}.$$ (7.23)

The denominators are the same as before, since

$$\exp(\log \lambda + \beta_{0t} + \tilde{\mathbf{x}}_i^T \tilde{\boldsymbol{\beta}}_t) + \exp(\log(1 - \lambda) + \beta_{0t} + \tilde{\mathbf{x}}_i^T \tilde{\boldsymbol{\beta}}_t) = \exp(\beta_{0t} + \tilde{\mathbf{x}}_i^T \tilde{\boldsymbol{\beta}}_t).$$

Thus the introduction of an irrelevant distinction changes a standard model into another standard model with two new states u and v in lieu of t, and these states have the same slope coefficients $\tilde{\boldsymbol{\beta}}_t$ as their parent state; only the intercepts differ. It follows that conversely in any standard multinomial model states with the same slope coefficients but different intercepts can be regarded as arbitrary subdivisions of a larger class and merged together. To test for pooling states is to test for the equality of all regressor coefficients apart from the intercept, or, for any two states u and v, to test the null hypothesis

$$\tilde{\boldsymbol{\beta}}_u = \tilde{\boldsymbol{\beta}}_v = \tilde{\boldsymbol{\beta}}_t.$$

This can be done by a straightforward likelihood ratio test. The test statistic of (3.11) of Section 3.1 is

$$\mathrm{LR} = 2[\log L(\hat{\boldsymbol{\theta}}_u) - \log L(\hat{\boldsymbol{\theta}}_r)],$$

or, for short

$$\mathrm{LR} = 2(\log L_u - \log L_r),$$

with u and r for the unrestricted and the restricted models respectively. Both models have $S + 1$ states, but in the restricted model two states have the same slope coefficients. If this restriction cannot be rejected, the two states are merged and we obtain a third model, the *pooled* model, with S states.

$\log L_u$ is available as a matter of course from fitting the original model; the question is to find a simple formula for $\log L_r$. To this end we order the observations by outcome and rewrite the loglikelihood accordingly, as in (3.4) of Section 3.1. Here we use

$$i \in \mathcal{A}_s$$

for the event that observation i has outcome s. In the present context $\mathcal{A}_u + \mathcal{A}_v = \mathcal{A}_t$. Setting the states u and v apart and renumbering the other states from 1 to $S - 1$, the unrestricted loglikelihood is

$$\log L_u = \sum_{i \in \mathcal{A}_u} \log P_{iu} + \sum_{i \in \mathcal{A}_v} \log P_{iv} + \sum_{s=1}^{S-1} \sum_{i \in \mathcal{A}_s} \log P_{is}.$$

Under the restriction of equal slopes, (7.22) applies, and this gives

$$\log L_r = \sum_{i \in \mathcal{A}_u} \log \lambda + \sum_{i \in \mathcal{A}_v} \log(1 - \lambda) + \sum_{i \in \mathcal{A}_u, \mathcal{A}_v} \log P_{it} + \sum_{s=1}^{S-1} \sum_{i \in \mathcal{A}_s} \log P_{is}.$$

This expression has four terms. The last two terms add up to the unconstrained loglikelihood of the *pooled* model with S states, say $\log L_p$. The first two terms depend on the parameter λ; if this is estimated by maximizing $\log L_r$ (as it should be), the result is

$$\hat{\lambda} = n_u / n_t$$

with n_s denoting the sample number of observations with outcome s. As a result, the first two terms of $\log L_r$ add up to

$$n_u \log n_u + n_v \log n_v - n_t \log n_t.$$

Altogether this gives

$$\log L_r = \log L_p + (n_u \log n_u + n_v \log n_v - n_t \log n_t).$$

The LR test statistic may therefore be obtained from two estimation runs, one of the original model with $S + 1$ states which gives $\log L_u$, and one of the pooled model with S states which gives $\log L_p$; upon adding the term in the numbers n_u, n_v and n_t this gives $\log L_r$. The formula is easily extended to more than two states; if state t is split into J states tj, the last formula becomes

$$\log L_r = \log L_p + \left(\sum_j n_{tj} \log n_{tj} - n_t \log n_t \right). \qquad (7.24)$$

As an illustration, we test whether the three categories of private car ownership in the analysis of the last section can be pooled, that is whether there are significant differences in their determination by the five covariates employed. The null hypothesis is that they are the same, and that the multinomial model can be reduced to the binary model with the same regressor variables of Section 3.7.

The restricted loglikelihood consists of two terms, viz. the loglikelihood from the pooled model and the contribution of the sample distribution of car ownership over the three subcategories. The first term is taken from Table 3.4 of Section 3.7:

$$\log L_p = -1351.39.$$

From the sample numbers of car–owning households at the beginning of Section 7.5 we find

$$\sum_j n_{tj} \log n_{tj} - n_t \log n_t = -1688.75$$

so that

$$\log L_r = -3040.14.$$

From Table 7.1 we have

$$\log L_u = -2874.90,$$

so that finally

$$\text{LR} = 330.48.$$

Since five covariates are involved and three parameter vectors have been constrained to equality, ten restrictions have been imposed, and this is the number of degrees of freedom of the chi-squared distribution. The restriction is soundly rejected, and the multinomial model is a significant improvement over the binary analysis.

8

Discrete choice or random utility models

This chapter gives an account of the theoretical and technical innovations in the econometrics of multinomial models that have been crowned by the Nobel prize for McFadden. Much of this work arose from applied research, primarily in transportation and marketing, and it has not widely spread beyond these fields. The firmly behavioural interpretation of the general logit model of discrete choice has been very fertile in bringing forth new generalizations and new specifications of the argument of the logit transformation. These techniques are however not necessarily linked to maximizing behaviour and they could be used with advantage in other fields.

8.1 The general logit model

The standard multinomial logit model is a special case of the *general* or *universal* logit model

$$P_{is} = Pl(V_{is}) = \frac{\exp V_{is}}{\sum_{t \in \mathcal{S}_i} \exp V_{it}}, \quad s \in \mathcal{S}_i. \tag{8.1}$$

The summation is over *sets* $\mathcal{S}_i$ of feasible or accessible states, which may vary from one i to another. The arguments V_{is} are further specified in various ways as functions of observed traits of observation i and of unknown parameters.

In practice there are two main classes of linear specifications of the V_{is}. The first is the standard model of (7.3) of Chapter 7:

$$P_{is} = Pl(\mathbf{x}_i^T \boldsymbol{\beta}_s^*), \quad \mathcal{S}_i = \mathcal{S} \text{ for all i.}$$

This mainstream form of multinomial logistic regression is obtained from

126

the general model by specifying the V_{is} as

$$V_{is} = \mathbf{x}_i^T \boldsymbol{\beta}_s^*. \tag{8.2}$$

The elements of $\mathbf{x}_i$, including the dummy unit constant, are *generic* covariates of observation i that have the same value for all states. In an analysis of the choice of transport modes s for a particular trip i they may reflect characteristics of the individual making the trip like gender, age and income, as well as other conditions like the time of day or the distance involved. The parameters $\boldsymbol{\beta}_s^*$ are *specific* for the transport modes or outcome states. The need to normalize these parameters with respect to a reference state, as in (7.4) of Section 7.2, is directly due to the form of this specification. This simple and robust specification is usually accompanied by the assumption of a single identical choice set of states $\mathcal{S}$ for all i.

The second major specification is the *conditional* logit model, where V_{is} is specified as

$$V_{is} = \tilde{\mathbf{z}}_{is}^T \tilde{\boldsymbol{\gamma}}.$$

In this pure version the vectors contain no unit constant and no intercept. Here the roles have been reversed: the covariates $\tilde{\mathbf{z}}_{is}$ are specific and vary with the states s (and possibly also with the observation i), and the slope coefficients are generic and the same for all states. If the s are again modes of transport, the elements of $\mathbf{z}_{is}$ describe properties like safety or comfort (the same for all i) and the duration and cost of a trip (varying with s but also with i). As the description of an option by the specific covariates is never complete, a state-specific intercept is often added to reflect the remaining unique qualities of each alternative. This gives

$$V_{is} = \beta_{0s} + \tilde{\mathbf{z}}_{is}^T \tilde{\boldsymbol{\gamma}}. \tag{8.3}$$

The unit constant is generic and the intercept is specific, and these β_{0s} must be normalized with respect to a reference state. In many applications other state-specific covariates are added as well, and the two specifications are combined into a single expression

$$V_{is} = \beta_{0s} + \tilde{\mathbf{x}}_i^T \tilde{\boldsymbol{\beta}}_s + \tilde{\mathbf{z}}_{is}^T \tilde{\boldsymbol{\gamma}} \tag{8.4}$$

with a reference state $s = 1$ where all β are zero. The conditional logit model is usually associated with more sophisticated studies than the standard model, and it is here that we find variation of the set $\mathcal{S}_i$ from one observation to another.

An important property of the general model is that for any pair (s, t) in the choice set $\mathcal{S}_i$ the logarithm of the ratio of any two probabilities $R(s, t)$ is given by

$$R_i(s, t) = \log(P_{is}/P_{it}) = V_{is} - V_{it}. \tag{8.5}$$

R_i, like the ratio of any two probabilities, depends exclusively on the characteristics of the two states concerned, and it is independent of the number and nature of all the other states that are simultaneously considered; in the conditional model they are not affected by changes in the values of alien covariates $\mathbf{z}_{iv}$. In consequence, the introduction of a new alternative (or the deletion of an existing option), or a change in the specific characteristics of a state, will alter the probabilities of all other outcomes in the same proportion, leaving their ratios unchanged. This property of the general logit model is known as *independence of irrelevant alternatives* or *IIA*. Since it also applies to highly relevant alternatives, it can lead to unacceptable results. The standard example is that of the red and blue buses versus private transport.† Consider the choice among several modes of transport, with s denoting private transport and t public transport, more specifically, travel by a red bus. This is the only available form of public transport. Suppose now that a new bus service is added to the choice set $\mathcal{S}_i$ that is almost identical to the existing service but uses blue buses. If the general logit model holds, (8.5) implies that the probability ratio for (t, s) and indeed all probability ratios from the original choice set are unchanged; the blue buses will therefore gain their share of the market by a proportional reduction of the probabilities of all previously existing transport modes. This is an unpalatable result, for in practice red bus traffic will suffer much more than other travel modes. The general logit model (and all its special forms) makes no allowance for this phenomenon.

The IIA property is due to the blind indifference of the model to any similarity or dissimilarity of the S states. In many applications this substantive assumption is clearly inappropriate. This defect cannot be remedied by an adjustment of the general logit model, but only by changing to a different probability model like the nested logit model or the multinomial probit model that are briefly discussed at the end of this chapter.

The standard multinomial and the conditional logit differ in the varia-

---

† This classic example is due to McFadden (1974). Debreu (1960) demonstrates the IIA by considering various recordings of the same concerto as alternatives to a live performance.

tion of the covariates and of the parameters. They can be combined, and with some ingenuity still other specifications of V_{is} may be designed. The broad dichotomy between standard and conditional models is a matter of practical considerations and of the observed regressor variation, not of profound theories about the process that generates the outcomes. In practice, however, the standard model is often treated as a mechanistic extension of the binary model, itself based on diverse ad hoc theoretical arguments, while the conditional model is linked specifically to the sophisticated theories of maximizing behaviour of *random utility* or *discrete choice*, which have a wide following in the fields of transportation and market research. We used this interpretation in discussing the IIA property, and we shall adhere to it throughout the present chapter. But there are no grounds for these ideological distinctions: discrete choice theory may very well be used in conjunction with the standard model, and the conditional logit (like other models that discrete choice theory has brought forth) is a valid statistical model in its own right without the particular connotation of random utility maximization.

The wider paradigm of discrete choice theory first arose in mathematical psychology, where the need to explain variations in repeated experimental measurements of individual preferences led to the notion of probabilistic choice. The seminal work is that of Thurstone (1927), and some ideas can even be traced all the way back to Fechner (1860). In the 1950s, theorists turned to the abstract mathematical representation of the choice process by the choice probabilities P_s of a given subject; for a survey of this work, see Luce and Suppes (1965). This led to probabilistic analogues of the preference relations of the classical theory of consumer behaviour: the deterministic relation 's is preferred to t' is replaced by 'given the choice between s and t, the probability that s is selected exceeds 0.5.' The next step is to search for equivalents of properties like the transitivity of preference relations, and to see whether probabilities with the requisite regularity properties can be linked to the maximization of an underlying utility function. At this stage, a distinction is in order between the *random utility model*, where the actual choice reflects the maximization of random utility indicators of the feasible options, and the *constant utility model* in which the utilities are determinate but the choice process is probabilistic.

In the context of constant utility, Luce (1959) imposed the *choice axiom*, whereby the conditional probability of selecting the state s out of a given subset is constant, regardless of the wider choice set to which this subset belongs. This is a mirror image of the IIA property. The axiom ensures that the probabilistic preference relation between two states holds regardless of what other states are considered, and once it is adopted the theory can be developed further without the restriction to a fixed and fully enumerated choice set, which may raise awkward problems. Luce also shows that this choice axiom implies the existence of a function ϕ_s such that

$$\Pr(s \text{ is chosen out of } \mathcal{S}) = \phi_s / \sum_{t \in \mathcal{S}} \phi_t,$$

where $\mathcal{S}$ is any set that contains s and at least one other choice. With minor additional assumptions Luce then proceeds to derive (8.1) from this strict utility model, with $\exp V_{is}$ taking the place of ϕ_s and $\mathcal{S}$ replaced by the feasible set $\mathcal{S}_i$.

The general logit model may also be derived from a random utility model. The discrete choice then reflects the maximization of (perceived) utility, which is a random attribute of feasible alternatives, and the model is driven by the distribution of these random variables. This very fruitful innovation of McFadden dates from the early 1970s; we shall reproduce his derivation in the next section.

8.2 McFadden's model of random utility maximization

In this derivation the general logit model reflects the maximization of the utility U_{is}, which is a random attribute of $\mathcal{S}_i$ alternatives. This process is conditional upon the given characteristics of observation i (or of individual i or experiment i), but in the present section this subscript is deleted. The random utility of outcome s is then defined as

$$U_s = V_s + \varepsilon_s, \tag{8.6}$$

where V_s is a systematic component and ε_s a random disturbance. Manski (1977) lists the origins of this random component: unobserved characteristics of the ith experiment, unobserved taste variation and similar imperfections. It may also accommodate a genuine indeterminacy of individual behaviour which calls for a probabilistic description, although this is strictly alien to deterministic utility maximization over random utilities.

With random utilities U_s and a feasible choice set $\mathcal{S}$, utility maximization implies choice probabilities

$$P_s = \Pr(U_s > U_t \text{ for all } t \neq s \in \mathcal{S}). \tag{8.7}$$

The P_s are thus determined by the V_s and by the stochastic specification of the ε_s. McFadden has established that P_s satisfies the general model of (8.1) if the disturbances ε_s are independently and identically distributed according to a type I extreme value distribution of standard form. We reproduce this derivation from McFadden (1974) and Domencich and McFadden (1975).

First, rewrite P_s as

$$P_s = \Pr(U_s > \breve{U}^s),$$

with

$$\breve{U}^s = \max(U_t,\ t \in \bar{S}),$$
$$\bar{S} = S - s.$$

In words, U_s must exceed $\breve{U}^s$, which is the largest of all other utilities.[†] The distribution functions of U_s and $\breve{U}^s$ are

$$F_{s1}(x) = \Pr(U_s \leq x)$$

and

$$F_{s2}(x) = \Pr(\breve{U}^s \leq x) = Pr(U_t \leq x,\ U_v \leq x,\ U_w \leq x, \ldots).$$

Since the ε_s are stochastically independent, so are the U_s (although they are not identically distributed as they have different parameters V_s). As a result

$$F_{s2}(x) = \prod_{t \in \bar{S}} F_{t1}(x).$$

Given proper analytical expressions for F_{s1} and F_{s2}, P_s can be obtained by the convolution theorem (see, for example, Mood et al., (1974, p. 186). By this theorem we have for any two independent random variables y and z

$$\Pr(y > z) = \int_{-\infty}^{+\infty} F_y'(t) F_z(t) \mathrm{d}t.$$

Here we have

$$P_s = \Pr(U_s > \breve{U}^s) = \int_{-\infty}^{+\infty} F_{s1}'(t) F_{s2}(t) \mathrm{d}t. \tag{8.8}$$

This completes the preliminaries.

---

† We ignore the possibility of a tie, which would arise if U_s were equal to $\breve{U}^s$. In the algebra that follows, strict and weak inequalities are treated in similar cavalier fashion.

It is now assumed that the disturbances ε_s are independent and identically distributed according to the *type I extreme value distribution in standard form*, also known as the *log Weibull* distribution and sometimes associated with *Gumbel*. This has the distribution function

$$F(x) = \exp[-\exp(-x)]. \tag{8.9}$$

In the present case

$$F_{s1}(x) = \Pr(U_s \le x) = \Pr[\varepsilon_s \le (x - V_s)]$$

gives

$$F_{s1}(x) = \exp[-\exp(V_s - x)]$$

for the distribution function of U_s. For $F_{s2}(x)$ we must take the product, or

$$F_{s2}(x) = \exp - \sum_{t \in \mathcal{S}} \exp(V_t - x)$$

$$= \exp[-\exp(\breve{V}_s - x)],$$

with

$$\breve{V}_s = -\log \sum_{t \in \mathcal{S}} \exp V_t.$$

Upon substituting these expressions into (8.9) we obtain

$$P_s(x) = \int_{-\infty}^{\infty} \exp V_s - t \exp[-\exp(V_s - t)] \exp[-\exp(\breve{V}_s - t)]\mathrm{d}t$$

$$= \exp V_s \int_{-\infty}^{\infty} \exp\{-t \exp\{-\exp[-t(\exp V_s + \exp \breve{V}_s)]\}\}\mathrm{d}t.$$

We now make use of

$$\frac{\mathrm{d}}{\mathrm{d}t} \exp[-A \exp(-t)] = A \exp\{-t \exp[-A \exp(-t)]\}$$

with

$$A = \exp V_s + \exp \breve{V}_s,$$

to write

$$P_s(x) = \frac{\exp V_s}{\exp V_s + \exp \breve{V}_s} \int_{-\infty}^{+\infty} \mathrm{d}\exp[-\exp(-tA)],$$

and obtain

$$P_s(x) = \exp V_s / (\exp V_s + \exp \breve{V}_s).$$

Finally we use (8.9) to find the expression we are looking for, namely (8.1)

$$P_s(x) = \exp V_s / \sum_t \exp V_t.$$

This completes the derivation of the general logit model from random utility considerations.

We have recorded these involved algebraic exercises to bring out the crucial assumption in the argument, which is that the ε_s are independent and identically distributed according to the standard type I distribution of (8.9). This is not a natural assumption. The stochastic independence of the ε_s across alternative choices is a strong restriction; it lies at the root of the IIA property. The ε_s must moreover follow a particular distribution with particular parameter values that are the same for all alternative choices. We briefly discuss what these assumptions imply.

In its general form the type I extreme value distribution function has two parameters, μ and λ, and its distribution function is

$$F(x) = \exp\{-\exp[-(x-\mu)/\lambda]\}. \tag{8.10}$$

Johnston and Kotz (1970, vol. 1, Ch. 21) review this distribution and its properties. There is no clear link of the present usage of the distribution with its derivation from the asymptotic behaviour of extreme sample values. If z has this distribution its mean and variance are

$$\mathrm{E}z = \mu + 0.5772\lambda,$$

$$\mathrm{var}\,z = 1.6449\lambda^2.$$

In the *standard* form of the distribution, μ is 0 and λ is 1, and the mean and standard deviation are 0.58 and 1.64. The standard density function is shown in Figure 8.1; it is not so very different in shape from a normal distribution. With $S = 2$ (and the specification of V_{is} as $\alpha_s + \beta_s X_i$) the present model reverts to the binary model of Chapter 2: the disturbance with the logistic density of (2.15) of Section 2.3 is the difference of two independent variates with the standard form of the extreme value distribution.

Whether the assumption of a common mean and variance with given values for all ε_s is an effective restriction depends on the specification of V_s. Suppose the true model is

$$U_s^* = V_s^* + \varepsilon_s^*$$

and the ε_s^* have an extreme value distribution with distinct parameters

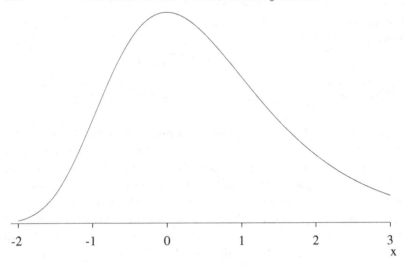

Fig. 8.1. Extreme value probability in standard form.

μ_s and λ_s for each state s. The desired *standard* disturbances ε_s are related to ε_s^* by

$$\varepsilon_s = (\varepsilon_s^* - \mu_s)/\lambda_s,$$

so that

$$U_s^* = V_s^* + \lambda_s \varepsilon_s + \mu_s.$$

Whether this can be rewritten like $U_s = V_s + \varepsilon_s$, as in (8.6), depends on the specification of V_s^* and its ability to absorb μ_s and λ_s. This operation is analogous to the treatment of the latent regression equation in Section 2.3, but the outcome is different. First, the mean μ_s can be absorbed as always by state-specific intercepts, as in (8.2) and (8.3). But the λ_s can *not* be accommodated by rescaling the slope coefficients of either specification, for if we were to divide these coefficients by different λ_s this would affect the ordering of the V_s and change the inequality (8.7).†
We must therefore maintain the assumption that all λ_s are equal, in other words that the disturbances have the same variance for all s. The slope coefficients (and the V_s and U_s) can then be normalized by a single rescaling factor so as to reduce their common λ to 1 and their common variance to 1.64.

† For the standard multinomial model this argument holds both before and after normalizing the parameters with respect to the reference state.

This assumption of a common variance of the disturbances for all s is an effective restriction. It is for example incompatible with the view that the disturbances take care of the effects of neglected variables, for these will vary across states. Whether the restriction is acceptable depends on the nature of the choice process, the definition of the alternatives and the nature of the observations. A consumer's choice of breakfast beverage can be a choice between tea and coffee, between blends, brands or places of purchase of coffee; the observations can refer to households or to individuals, they can be restricted to a certain locality or a certain type of shop, and they can consist of repeated observations for the same consumers. All these aspects have a bearing on the distribution of the random elements in the choice process. For a review of the subtle consideration of these matters in market research, and of the various models that meet the ensuing complications, see Baltas and Doyle (2001).

So much for the algebra and the implicit assumptions of this operational discrete choice theory. While it rests on a narrow basis, it has been very fertile in applied research like policy analyses of transport and travel demand and marketing research. While it may equally well lead to the standard multinomial as to the conditional logit model, its exceptional power has mainly made itself felt with the latter specification. It is here that the interpretation of V_s as the systematic utility of one option relative to the others is fully exploited. Some of its advocates believe that discrete choice theory is also superior to other derivations of the general logit model because it permits a *structural* analysis of observed choice behaviour, as the probability model is directly linked to underlying utility-maximizing behaviour. Forceful statements of this doctrine are found in the writings of McFadden (1976, 2001) and Manski (1977). The theory does indeed provide an instance of that rare and much sought after prize, economic theory dictating the form of an empirically valid mathematical function; but to make the theory work some fairly arbitrary assumptions have been necessary. It is probably a greater achievement of the theory that it has provided an effective vehicle for the conditional logit model as well as for several more advanced models that are free from the IIA property.

8.3 The conditional logit model

In the conditional logit model (8.3)

$$V_{is} = \beta_{0s} + \tilde{\mathbf{z}}_{is}^T \tilde{\gamma}$$

the elements of $\tilde{\mathbf{z}}_{is}$ represent specific properties of the various options s that may or may not vary with the conditions of observation i. For modes of transport they reflect safety or speed (which do not vary with i) or costs or time (which do), for college choice they represent tuition costs (the same for all i) and distance from the freshman's home (varies with i). There are S vectors $\mathbf{z}_{is}$ of k elements for each observation, but there are only k generic parameters in $\tilde{\gamma}$, and another $S-1$ parameters in the β_{0s}. In contrast to the generic β of the standard multinomial model, γ does not need to be normalized in the interest of identification.

An early example that conveys the flavour of such analyses is the study of travel behaviour of Washington commuters by Lerman and Ben-Akiva (1976). They distinguish five travel modes, defined as combinations of household car ownership and car use, such as 'two cars, but not used for travel to work'. The utility of each mode depends on variables like travel time, a combined variable for travel costs in relation to income, and the ratio of cars to licensed drivers in the household, with different definitions for car-to-work travel modes and others. In a similar vein, Gaudry et al. (1989) distinguish nine travel modes for office workers in Santiago de Chile, and employ covariates like several components of the time for the trip (walking time, waiting time, in-vehicle time) and the cost as a fraction of the traveller's income.

The assumption of a single $\tilde{\gamma}$ for all modes is only justified if it is believed that aspects like costs and travel time, if properly measured, affect the utility of all modes for a given trip in the same way. The terms of $\mathbf{z}_{is}^T\tilde{\gamma}$ then give the contribution of each aspect to the overall utility V_{is} of a given mode, and the elements of $\tilde{\gamma}$ reflect the relative weight of these aspects. Cost and speed will affect utility with opposite sign: the ratio of their coefficients then marks the trade-off between these aspects at constant utility, and shows how much a traveller is willing to pay for a given saving in time. Apart from this useful interpretation of the parameters, the model also permits extrapolation to a new mode of travel v with known characteristics $\mathbf{z}_{iv}$ for each i. Once $\tilde{\gamma}$ has been estimated, we can calculate V_{iv} and insert this in the model. This is useful when the introduction of a new option or product with known properties is contemplated. The analysis can even be extended to a new product with characteristics that have not yet been experienced by the public, making use of survey data about *stated preferences* (as opposed to the revealed preferences of observed behaviour). In these interviews respondents are asked to give their opinion on hypothetical products that are described in great detail. This technique was already used by Adam (1958) for probit analyses of the willingness to pay for particular products; for a much more sophisticated analysis, employing an ordered logit model with both specific and generic covariates, see Beggs et al. (1981). But it will be clear that this is also precisely the type of application where the limitations of the IIA property are most keenly felt.

In a pure conditional logit, systematic utility V_{is} is completely described by the $\mathbf{z}_{is}$. This resembles the view of all commodities as bundles of properties that make up their utility, as in the consumer demand theory of Lancaster (1971). In practice, the covariates are an incomplete description, and a state-specific intercept is added, as we have done above. This can also be done by introducing state-specific unit variables for each of $S - 1$ states as part of the $\mathbf{z}_{is}$, with $S - 1$ additional coefficients in $\boldsymbol{\gamma}$; this technique is illustrated in Table 8.2 below. – State-specific intercepts complicate the prediction of the utility of a new option v, for it will be necessary to put a value on β_v. And this becomes worse if other generic covariates have been added as well, as in (8.4).

The properties of the conditional logit model (with or without generic regressors) are easily derived. The derivative of the probabilities with respect to the jth element of $\mathbf{z}_{is}$ or $\mathbf{z}_{it}$ under given conditions i are

$$\begin{aligned} \partial P_{is}/\partial Z_{is,j} &= \gamma_j P_{is}(1 - P_{is}), \\ \partial P_{is}/\partial Z_{it,j} &= -\gamma_j P_{is} P_{it}. \end{aligned} \tag{8.11}$$

The derivatives can be evaluated at the sample mean, with the sample frequencies substituted for the probabilities. As an example, s can be private transport and t public transport, $Z_{is,j}$ and $Z_{it,j}$ their costs for a particular trip i, and γ_j the coefficient that reflects the (dis)utility of expenditure on fares. In contrast to the standard multinomial model, the probabilities vary monotonically with any Z_j, with the sign depending on γ_j. Note that the effect on the probability of state s of changes in its own properties and in the same properties of other states are of opposite sign: if P_{is} declines upon a cost increase of mode s, it increases if the cost of other modes is cut.

By (8.11) the cross-effects exhibit symmetry, or

$$\partial P_{is}/\partial Z_{it,j} = \partial P_{it}/\partial Z_{is,j}. \tag{8.12}$$

It is easily verified that

$$\sum_s \partial P_{is}/\partial Z_{it,j} = 0,$$

as well as

$$\sum_t \partial P_{is}/\partial Z_{it,j} = 0.$$

The first expression states the obvious (that the probabilities sum to 1), and the second shows that the probabilities remain the same upon the same change in $Z_{it,j}$ for all t, as they should. For if all $Z_{it,j}$ change by the same amount, all utilities V_{is} also change by the same amount, and the preferred choice at observation i remains the same.

Quasi-elasticities are obtained as before by multiplying the derivatives by the relevant values of $Z_{is,j}$ or $Z_{it,j}$, or

$$\eta_{ss} = \gamma_j Z_{is,j} P_{is}(1 - P_{is}),$$

$$\eta_{st} = -\gamma_j Z_{it,j} P_{it} P_{is}.$$

For the estimation of the conditional logit model we once more rely on the iterative scheme of (7.9) of Section 7.3. All we need to complete the score vector $\mathbf{q}$ and the Hessian $\mathbf{H}$ are the derivatives of the probabilities with respect to the parameters, which are then inserted in the general expressions

$$q_{ij} = \sum_{s=1}^{S} \frac{y_{is}}{P_{is}} \frac{\partial P_{is}}{\partial \theta_j}$$

and

$$H_{ijh} = \sum_{s=1}^{S} \frac{1}{P_{is}} \frac{\partial P_{is}}{\partial \theta_j} \frac{\partial P_{is}}{\partial \theta_h}.$$

The derivatives with respect to the elements of γ are

$$\partial P_{is}/\partial \gamma_j = P_{is} \left(Z_{is,j} - \sum_t Z_{it,j} P_{it} \right).$$

Upon defining a weighted average of the $Z_{is,j}$ for observation i over all modes,

$$\bar{Z}_{ij} = \sum_t Z_{it,j} P_{it},$$

this can be rewritten as

$$\partial P_{is}/\partial \gamma_j = P_{is} \left(Z_{is,j} - \bar{Z}_{ij} \right). \tag{8.13}$$

Substitution in the score gives

$$q_{i\gamma_j} = \sum_s Y_{is} \left(Z_{is,j} - \bar{Z}_{ij} \right).$$

The estimates enter into this expression via the weights P_{is} of $\bar{Z}_{ij}$. Note that the summation over s yields only a single term, as all but one of the Y_{is} are zero. As before, the maximum likelihood estimates will satisfy

$$\sum_i q_{i\gamma_j} = 0$$

but this does not lead to simple side relations, as in the case of the standard model. The equality of the means, for example, is not ensured, unless there are state-specific intercepts; in that case the argument of (7.19) of Section 7.4 applies. In a pure conditional model, it will only hold approximately in large samples because of an entirely different argument, namely (4.6) of Section 4.4.

For an element of $H_{i\gamma}$ we obtain,

$$H_{ijh} = \sum_s P_{is} \left(Z_{is,j} - \bar{Z}_{ij} \right) \left(Z_{is,h} - \bar{Z}_{ih} \right),$$

which looks like a weighted moment. The weights P_{is} (which also enter into the $\bar{Z}$) vary of course with the parameters that are being estimated, as is usual in iterative schemes.

If we wish to estimate the fuller model (8.4), $\mathbf{q}_i$ and $\mathbf{H}_i$ can be partitioned as

$$\mathbf{q}_i = \begin{bmatrix} \mathbf{q}_{i\beta} \\ \mathbf{q}_{i\gamma} \end{bmatrix},$$

and

$$\mathbf{H}_i = \begin{bmatrix} \mathbf{H}_{i\beta\beta} & \mathbf{H}_{i\gamma\beta} \\ \mathbf{H}_{i\beta\gamma} & \mathbf{H}_{i\gamma\gamma} \end{bmatrix}.$$

The terms in β have been given in Section 7.4, and the terms in γ have just been derived; for the off-diagonal matrices of $\mathbf{H}$, the derivatives of (7.15) of Section 7.4 and of (8.13) must be combined, but this presents no particular difficulties.

If the choice set varies between observations, summation over s for each i should take place over a set $\mathcal{S}_i$ that varies from one observation to another. Differences in the choice set will also affect $\mathbf{q}_{i\beta}$ and the elements of $\mathbf{H}_i$ that involve the β parameters, as the absence of a state implies the absence of a subset of β, and zero elements in the corresponding places. But this is a matter of careful programming which raises no questions of principle.

8.4 Choice of a mode of payment

We illustrate the conditional model for the choice of a mode of payment by Dutch consumers in 1987.† The example is outdated, but it is instructive insofar as both the standard multinomial and the conditional logit are applied with almost the same covariates. The major difference is that the standard model describes while the conditional model explains.

Point-of-sale payments by anonymous clients in shops and restaurants demand immediate and secure settlement. At the time of the study the only acceptable means of payment were cash and guaranteed cheques, available in three types with slightly different properties. The analysis was intended to contribute to improved conditions of cheque use and the design of new modes of payment; but it was soon superseded by technological progress, which permitted the widespread installation of point-of-sale terminals for electronic transfer. This has completely replaced cheques.

The Dutch Intomart household expenditure panel recorded the mode of payment (cash or cheque of a particular type) of each item of expenditure, the person making the purchase and the place of payment (shop, bank, home). We use data on 2161 point-of-sale payments in cash or by guaranteed cheque from a random sample of payments by 1000 households. There are three types of cheque, all guaranteeing payment to the recipient up to a certain limit, provided the payer has shown a bank card; they are here distinguished by their colour. The *green* cheque, issued by the commercial banks, is guaranteed for sums up to fl. 100, roughly the equivalent of $50; the *orange* cheque of the postal giro system has an upper limit of fl. 200; and the *blue* Eurocheque, issued by the banks along with the green cheque, has an upper limit of fl. 300. Access to the blue cheque, which is also valid in other European countries, is somewhat restricted and carries a small charge. All limits apply to a single cheque, and larger sums can be paid by writing several cheques. Very nearly all Dutch households (and all households in the sample) have an account with the postal giro system and/or with one or more banks, with the giro most widely prevalent among all classes of society. In principle, all three types of cheque are available to all households and in practice very nearly so. With cash this makes four modes of payment and we assume that these are accessible to all payers.

---

† This example is taken from a study commissioned by the Postbank and earlier
  reported in Mot et al. (1989) and Mot and Cramer (1992)

Table 8.1. *Point-of-sale payments, Dutch households, 1987.*

	cash	guaranteed cheque		
		green	orange	blue
number of observations	1899	32	131	99
	column percentage			
less than fl. 100	97.2	78.1	73.6	67.7
fl. 100 – fl. 200	2.2	18.8	17.6	25.3
fl. 200 – fl. 300	0.4	3.2	4.6	3.0
over fl. 300	0.3	0.0	2.3	4.0

Table 8.1 shows that the vast majority of point-of-sale payments are made in cash, and that only the larger payments are made by cheque. The object of the analysis is to find out why this is so. We employ three logit models to answer this question. *Model A* is the standard multinomial model of (7.5) for four modes of payment, with the logarithm of the amount paid of LSUM as the single regressor apart from the unit constant. *Model B* is the pure conditional logit of (8.2) with two specific covariates Z_{is} called RISK and INCONVENIENCE, two aspects of the intangible transaction costs; they have generic coefficients. In *model C* specific intercepts are added along the lines of (8.3).

Table 8.2 sets out the variables of the three models. For *Model A* these are the unit constant and LSUM; the use of the logarithm is a purely empirical device, adopted because it gives a better fit. For *models B* and *C* we must find operational measurements of the risk and inconvenience associated with payment by a particular mode. By risk we mean the dangers of theft or loss associated with a payment. For cash payments this is just the sum paid, for this is the amount of cash the payer must carry; once more the logarithm of this amount is taken, but now this transformation reflects the way the loss is felt. Thus CASH RISK is measured once more by LSUM. For the guaranteed cheques, risk is the perceived cost of losing cheques and/or the accompanying bank card, for the payer must carry both to make payments. The issuing banks usually limit these costs to a fixed amount, depending on the circumstances of the loss or theft, but the public are not always aware of the precise rules. We assume that the perceived CHEQUE RISK is constant and the same

Table 8.2. *Choice of a mode of payment: three models and their variables.*

	cash	guaranteed cheque		
		green	orange	blue
Model A				
X_{i1} CONSTANT		1		
X_{i2} LSUM		log(amount)		
Model B				
Z_{i1} CASH RISK	log(amount)	0	0	0
Z_{i2} CHEQUE RISK	0	1	1	1
Z_{i3} INCONVENIENCE	0	NRR	NRR	NRR
Model B				
Z_{i1} CONSTANT1	0	1	0	0
Z_{i2} CONSTANT2	0	0	1	0
Z_{i3} CONSTANT3	0	0	0	1
Z_{i4} CASH RISK	log(amount)	0	0	0
Z_{i5} INCONVENIENCE	0	NRR	NRR	NRR

NRR *is the number of cheques required for the payment.*

for all three cheques: the covariate is a unit constant, and its coefficient will measure the monetary value. As for INCONVENIENCE or loss of time, this is measured by the number of cheques that the payment requires. For currency it is zero, regardless of the amount paid; for cheques it is the number of cheques that must be filled in and signed. This takes time; payers are also reluctant to part with cheques because they receive only a limited number at a time from their bank. The number required or NRR is one more than the integer part of the quotient of the amount paid and the cheque limit; this varies with the sum paid and also with the type of cheque, since these have different limits.

Model B is a pure conditional logit, with RISK and INCONVENIENCE together completely determining the utility of each mode of payment; model C allows for other, unobserved differences by specific intercepts, with cash the reference state. The generic unit constant of (8.3) has here been replaced by three specific unit covariates for the three cheques, each with its own (generic) coefficient. As these three covariates together are identical to CHEQUE RISK of model B, its coefficient can no longer be

Table 8.3. *Estimates of three models of*
mode-of-payment.

Model A : Standard logit		
6 parameters, $\log L = -826.55$		
quasi − elasticities with respect to amount paid :		
CASH	−0.18	(16.8)
GREEN CHEQUE	0.02	(6.6)
ORANGE CHEQUE	0.08	(13.1)
BLUE CHEQUE	0.07	(12.3)
Model B : Pure conditional logit		
3 parameters, $\log L = -839.12$		
coefficients of CASH RISK	−1.86	(18.0)
CHEQUE RISK	−8.44	(20.4)
INCONVENIENCE	−1.17	(5.0)
Model C : Conditional logit with intercepts		
5 parameters, $\log L = -816.55$		
coefficients of CASH RISK	−1.79	(17.4)
INCONVENIENCE	−0.91	(4.0)
intercepts GREEN CHEQUE	−9.25	(21.0)
ORANGE CHEQUE	−8.04	(19.5)
BLUE CHEQUE	−8.39	(20.2)

Absolute values of $t-ratios$ *in brackets*;
null loglikelihood − 1052.68.

estimated separately; in the discussion it is equated to the mean of the cheque intercepts, and the deviations from this mean are interpreted as the intrinsic utility of the three cheque modes relative to one another.

The estimates for the three models (with t-values in brackets) are shown in Table 8.3. Model A confirms what was already apparent from Table 8.1 as well as from casual observation, namely that small amounts are paid in cash and only larger sums by cheque. The quasi-elasticities are quite small, but then the amount paid varies from one payment to another by multiples, not by a few per cent. The very small elasticity of green cheques is probably associated with their limited face value.

These results are sensible but they throw no light on the question *why* cheques are generally reserved for the larger payments. The two conditional models attempt an answer to this question. In model B, the coefficient of −1.86 corresponds at the sample mean frequency to a derivative of P_{CASH} with respect to CASH RISK of $-1.86 \times 0.85 \times 0.15 =$ -0.23, and this is a quasi-elasticity since cash risk is the logarithm of the amount at stake. This quasi-elasticity applies with the properties of cheques held constant, and it is not surprising that it is larger than the quasi-elasticity in respect of the amount paid of Model A, even though they refer to the same variable. The risk of using cheques is equivalent to a cash risk of $\exp(-8.442/-1.86) =$ fl. 93.50, and this is a sensible result.

Model C yields equally precise estimates as A and B; the fit is significantly better than that of B, with an increase in $\log L$ of 22.57 for two additional parameters, and $\log L$ is also higher than in model A in spite of a smaller number of parameters (but these two models are not nested). The coefficients of CASH RISK and of INCONVENIENCE are not substantially different from model B. If we identify CHEQUE RISK with the weighted mean of the cheque intercepts, its coefficient is -8.32, near enough to the earlier estimate of -8.44. The specific utility of the three cheques relative to one another is then represented by the deviation of their intercepts from this weighted mean, or

green cheque -0.93,
orange cheque 0.28,
blue cheque -0.07.

Allowing for inconvenience and with the assumption of a common cheque risk, orange cheques are by far the most popular of the three, with blue cheques a good second and green cheques at a considerable distance. This may reflect the popularity of the postal giro system.

The major advantage of the conditional logit model is that it permits prediction of the likely effect of policy changes, like changes of the cheque risk or of the guaranteed limits (and hence of INCONVENIENCE). The variables of Table 8.2 or the estimated coefficients are changed accordingly, the probabilities of the four modes are recalculated for all payments in the sample, and then summed to give the new incidence of the various modes of payment. But we do not report the outcome of such exercises here.

8.5 Models with correlated disturbances

Several models have been put forward to break free of the restrictive IIA property within the random utility framework with the outcome determined by maximizing random utilities

$$U_{is} = V_{is} + \varepsilon_{is}.$$

Since the independence of the disturbances ε_s is at the root of the IIA property, the natural solution is to introduce a joint S-dimensional distribution of the ε_s that allows for their correlation. But without independence the easy algebra of the convolution theorem of (8.8) no longer applies, and formidable analytical obstacles arise over the inequality (8.7). Moreover the correlations may bring up to $S(S-1)/2$ additional parameters into play, and it is desirable to reduce this number by imposing a certain structure on the correlation matrix.

We shall briefly discuss two models of this kind. The first is based on a direct generalization of the extreme value distribution of (8.9), namely Gumbel's Type B extreme value distribution usually defined in the literature for the case $S = 2$; its distribution function is then

$$F(x, z) = \exp\{-[\exp(-x/\rho) + \exp(-y/\rho)]^\rho\}; \qquad (8.14)$$

see Johnson and Kotz (1972, p. 256) or Amemiya (1985, p. 306). If the distribution is generalized to more than two variates, the same correlation ρ holds for all pairs. This parameter reflects the dependence of the variates; it must lie between 0 and 1, and the correlation between any pair of variates is $1 - \rho^2$. In spite of the notation (which suggests the reverse), the limiting case $\rho = 1$ corresponds to independence, with all variates having the standard Type I extreme value distribution of (8.9); this brings us back to the original model and the IIA property.† This opens the way to a test of the IIA property by testing the nested parametric hypothesis $\rho = 1$ by the usual repertoire listed in Section 4.1. Hausman and McFadden (1984) develop such a test for a conditional logit as a restricted case of a nested logit (to be discussed presently), along with another test which is based on first principles of choice behaviour under IIA. If the property holds, the probability ratios of the remaining options are unchanged if one state is eliminated; one may test whether the parameter estimates from the full model and the reduced model differ significantly. The difficult part is the derivation of the covariance matrix of the differences between the two estimate vectors.

---

† The other extreme of $\rho = 0$ or perfect correlation implies identical disturbances, but not identical utilities, for the systematic components will usually differ. With identical utilities choice would be indeterminate; with identical disturbances but different systematic components one alternative is always preferred, and the entire stochastic framework of choice breaks down.

Since the assumption of the same correlation for all pairs of alternatives is only a little less restrictive than independence, flexibility is introduced in the *nested* logit model by imposing a hierarchical structure of the choice process which looks like a tree with a succession of branches. The various states are classified along these branches into subgroups or clusters of similar alternatives. Within each cluster the random disturbances are correlated and between clusters they are independent. This is the *nested logit model* of McFadden (1981), also discussed at some length by Amemiya (1985).

In the simplest case there are three states, with two related or similar alternatives forming one cluster and a third state that is independent of both. The stock example is once more the case of private driving and two public transport services or of the car versus red and blue bus services as transport modes for a given trip. The two bus services form a correlated cluster; note that they must differ in other aspects than the colour of their vehicles alone, for if they have the same systematic component V_{is} the analysis will break down. The car disturbance ε_1 has the Type I extreme value distribution of (8.9),

$$F(x) = \exp\left[-\exp(-x)\right]$$

and the other two disturbances ε_2 and ε_3 have the joint distribution function of (8.14). Amemiya (1985, pp. 300-306) derives the probabilities for the three alternatives from these distributional assumptions as

$$P_1 = \frac{\exp(V_1)}{\exp(V_1) + [\exp(V_2/\rho) + \exp(V_3/\rho)]^\rho},$$

$$P_2 = (1 - P_1)\frac{\exp(V_2/\rho)}{\exp(V_2/\rho) + \exp(V_3/\rho)},$$

and

$$P_3 = (1 - P_1)\frac{\exp(V_3/\rho)}{\exp(V_2/\rho) + \exp(V_3/\rho)}.$$

This suggests that the parameters of the V_s can be estimated in two steps. The first is restricted to the alternatives in the cluster, here the numbers 2 and 3; within this group, a straightforward conditional logit estimation will give estimates of the parameters of V_2 and V_3. If these are specified as $V_s = \mathbf{z}_{is}^T \boldsymbol{\gamma}$ the estimates refer to $\boldsymbol{\gamma}/\rho$. These estimates can then serve to construct estimated values of the denominators of P_1 in a subsequent binomial analysis at cluster level. With some ingenuity the estimation is thus reduced to routine procedures, although the variances of the estimates require separate further computation. But the nested model may also be estimated by classical maximum likelihood methods. Some of the advanced program packages provide ready routines.

In comparison with the general logit of (8.1), the nested version introduces greater flexibility to the tune of one additional parameter for each separate cluster. A full description of S random utilities calls for a much larger number. For utility maximization one state can be taken as the reference state, with all utilities measured relative to it; this is often identified with the preferred choice, so that this has utility zero and all alternatives have negative utilities. There remain $S - 1$ random utilities, with their means determined by the systematic component V_s; in the covariance matrix, one variance is set at 1 to determine the scale of the slope coefficients.† Then there are $S(S - 1)/2$ correlations (possibly with some constraints). The full richness of this specification is recognized in the *multinomial probit* model.

In this model the additive specification $U_{is} = V_{is} + \varepsilon_{is}$ is completed by the assumption

$$\varepsilon_s \sim N(\mathbf{0}, \mathbf{\Omega}^*).$$

Subsequently the dimension is reduced by 1 by taking out a reference state, and one (diagonal) element of the covariance matrix $\mathbf{\Omega}^*$ is equated to 1. The disturbances of the differential utilities still have a normal distribution (except for the reference state). In the case $S = 2$ this reduces to the standard normal distribution of the binary probit model of Section 2.3. As in (8.7), utility maximization implies

$$P_s = \Pr(U_s > U_t \text{ for all } t \neq s).$$

With a joint density $f(\cdot)$ of the U_s, the probability (say) P_1 is given by the multiple integral

$$P_1 = \int_{-\infty}^{\infty} \int_{-\infty}^{U_{s-1}} \ldots \int_{-\infty}^{U_1} f(U_1, U_2, \ldots, U_s) dU_1 dU_2 \ldots dU_s. \quad (8.15)$$

Estimation requires that this probability is expressed in terms of the parameters of the model, viz. the parameters of the systematic component and the parameters of the normal density. But the normal distribution does not permit of an analytical solution, and the integral must be evaluated by numerical methods. For the univariate and, to a lesser extent, for the bivariate distribution the difficulties are not insuperable, but for

---

† If the disturbances are identically distributed, as in Section 8.2, all variances are equal and this takes care of a further $S - 2$ parameters.

$S > 3$ the problem rapidly becomes intractable. Note that in the iterative schemes of maximum likelihood estimation the probabilities P_s must at each iteration be evaluated anew for all observations. Early analyses of the multinomial probit by Hausman and Wise (1978) and Daganzo (1979) are therefore limited to the case $S = 3$.

One way of solving this problem is by *simulating* the probabilities instead of evaluating the integral. The simplest form is to take a sample of random drawings from the multidimensional density and establishing the frequency of the outcomes that have U_s as the largest of all U_t. This idea was mooted by Lerman and Manski (1981) and later taken up independently by Pakes and Pollard (1989) and McFadden (1989). Since then the method has been further refined and improved by ingenious techniques which permit valid working probabilities (and sometimes also of their derivatives) from a small number of drawings from each distribution. For examples see Börsch-Supan and Hajivassiliou (1993) or Hajivassiliou et al. (1996). The illustrative simulations in these articles deal with five states, but in the transportation and market research literature examples are found with up to nine states, as in Gaudry et al. (1989). Problems of such larger dimension naturally arise in the analysis of panel data, where the stochastic structure is further complicated by autoregressive elements. Such generalizations are far removed from the initial moves to break free from the IIA property of the general logit model.

9
The origins and development of the logit model

The present practices of logistic regression, case–control studies and discrete choice analyses have separate and distinct roots, often spreading back to achievements of the 19th century. The development of these techniques has been governed equally by the immediate needs of such diverse disciplines as biology, epidemiology and the social sciences and by the personal histories of individual scholars. The present account is limited to the origins of the logistic function and its adoption in bioassay and in economics, followed by a brief survey of alternative routes to the same model.†

9.1 The origins of the logistic function

The sigmoid logistic function

$$W = \frac{\exp(\alpha + \beta t)}{1 + \exp(\alpha + \beta t)}\, \Omega$$

or a related form, embellished by additional parameters, was introduced in the 19th century for the description of population growth and of the course of autocatalytic chemical reactions. In either case we consider the time path of a quantity $W(t)$ and its growth rate

$$\dot{W}(t) = \mathrm{d}W(t)/\mathrm{d}t. \tag{9.1}$$

The simplest assumption is that $\dot{W}(t)$ is proportional to $W(t)$:

$$\dot{W}(t) = \alpha W(t), \quad \alpha = \dot{W}(t)/W(t),$$

---

† An updated paper giving fuller biographical and bibliographical sources can be found at http://publishing.cambridge.org/resources/0521815886/.

with α the constant relative growth rate. This simple differential equation leads of course to exponential growth,

$$W(t) = A \exp \alpha t,$$

where A is sometimes replaced by the initial value $W(0)$. This is a reasonable model for unopposed population growth in a young country like the United States in its early years; it can be argued that it lies at the basis of Malthus' contention of 1798 that a human population, left to itself, would increase 'in geometric progression'.† But Alphonse Quetelet (1795–1874), the formidable Belgian astronomer turned statistician, who took a great interest in vital statistics, was well aware that indiscriminate extrapolation of exponential growth must lead to impossible values. He experimented with various adjustments and then asked his pupil, the mathematician Pierre-François Verhulst (1804–1849), to look into the problem.

Verhulst added an extra term to the differential equation (9.1) as in

$$\dot{W}(t) = \beta W(t)[\Omega - W(t)], \tag{9.2}$$

where Ω denotes the upper limit or *saturation level* of W. We may express $W(t)$ as a fraction of Ω, say $P(t) = W(t)/\Omega$. (In this context P denotes a proportion, not a probability as in the earlier chapters.) This gives

$$\dot{P}(t) = \beta P(t)[1 - P(t)]. \tag{9.3}$$

As we know from (2.3) of Section 2.1, the solution of this differential equation is

$$P(t) = \exp(\alpha + \beta t)/[1 + \exp(\alpha + \beta t)], \tag{9.4}$$

which Verhulst named the logistic function. He published his findings between 1838 and 1847 in three papers, first in the *Correspondance Mathématique et Physique* edited by Quetelet and then in the Proceedings of the Belgian Royal Academy (Verhulst 1838, 1845, 1849). In the second paper he fitted (9.2) to the Belgian population and arrived at an estimate of the upper limit Ω of 6.6 million, which is far surpassed by the present population of 10.2 million in 1998.

The discovery of Verhulst was repeated by Pearl and Reed (1920). At the time Raymond Pearl (1879–1940) had just been appointed Director of the Department of Biometry and Vital Statistics at Johns Hopkins

---

† Exponential growth is still with us in economic analyses. It also played a major part in the *Report to the Club of Rome* of Meadows et al. (1972).

University, and Lowell J. Reed (1886–1966) was his deputy (and his successor when Pearl was promoted to Professor of Biology). Pearl was trained as a biologist, and acquired his statistics as a young man by spending a year with Karl Pearson in London. He became a prodigious investigator and a prolific writer on a wide variety of phenomena like longevity, fertility, contraception, and the effects of alcohol and tobacco consumption on health, all subsumed under the heading of human biology. During World War I Pearl worked in the US Food Administration, and this may account for his preoccupation with the food needs of a growing population in the 1920 paper. At that time, Pearl and Reed were unaware of Verhulst's work (though not of the curves for autocatalytic reactions discussed presently), and they arrived independently at a close variant of the logistic curve. They fitted this to the US population census figures from 1790 to 1910 and obtained an estimate of Ω of 197 million, which compares badly with the present value of 270 million in 1998. In spite of many other interests, Pearl and his collaborators in the next twenty years went on to apply the logistic growth curve to almost any living population from banana flies to the human population of the French colonies in North Africa as well as to the growth of melons – see Pearl (1927), Pearl et al. (1928) and Pearl et al. (1940) for examples. Reed, who had been trained as a mathematician, made a quiet career in biostatistics. He excelled as a teacher and as an administrator; after his retirement he was brought back to serve as President of Johns Hopkins. Among his papers in the aftermath of the 1920 paper with Pearl is an application of the logistic curve to autocatalytic reactions (Reed and Berkson 1929). We shall hear more about his co-author in the next section.

In 1918, Du Pasquier, Professor of Mathematics at Neuchâtel, Switzerland, published a brief survey of mathematical representations of population growth in which he names Verhulst (Du Pasquier 1918). There are some indications that it was Du Pasquier who informed Pearl and Reed of Verhulst's work. The first time these authors acknowledged Verhulst's priority is in a footnote in Pearl (1922); this is followed by a fuller reference in Pearl and Reed (1923). Some years later, Yule (1925), who says he owes the reference to Pearl, treats Verhulst much more handsomely and devotes an appendix to the writings of Quetelet and Verhulst on population growth. Yule also adopts the term 'logistic' from Verhulst. It would take until 1933 for Miner (a collaborator of Pearl) to write an article about Verhulst, largely a translation of an obituary by Quetelet with the addition of an extract from the published reminiscences of Queen Hortense de Beauharnais.

There is another early root of the logistic function in chemistry, where it was employed (again with some variations) to describe the course of autocatalytic reactions. These are chemical reactions where the product itself acts as a catalyst for the process. This leads naturally to a differential equation like (9.2) and hence to functions like the logistic for the time path of the amount of the reaction product. A review of the application of logistic curves to a number of such processes by Reed and Berkson (1929) quotes work of the German Professor of Chemistry Wilhelm Ostwald of 1883. Authors like Yule (1925) and Wilson (1925) were well aware of this strand of the literature.

The basic idea of logistic growth is simple and effective, and it is used to this day to model population growth and market penetration of new products and technologies. The introduction of mobile telephones is an autocatalytic process, and so is the spread of new products and techniques in industry.

9.2 The invention of probit and the advent of logit

The invention of the probit model is usually credited to Gaddum (1933) and Bliss (1934a,b), but one look at the historical section of Finney (1971) or indeed at Gaddum's paper and his references will show that this is too simple. Finney traces the roots of the method and in particular the transformation of frequencies to equivalent normal deviates to the German scholar Fechner (1801–1887). Stigler (1986) recounts how Fechner was drawn to study human responses to external stimuli by experimental tests of the ability to distinguish differences in weight. He was the first to recognize that response to an identical stimulus is not uniform, and to transform the observed differences to equivalent normal deviates. The historical sketches of Finney (1971, Ch. 3.6) and Aitchison and Brown (1957, Ch. 1.2) record a long line of largely independent rediscoveries of this approach that spans the seventy years from Fechner to the early 1930s. Gaddum and Bliss attach more importance to the logarithmic transformation of the stimulus than to the assumption of a normal distribution of the response threshold, which they regard as commonplace. Their publications contain no major innovations, but they mark the emergence of a standard paradigm of bio-assay, with Gaddum's report particularly effective in Britain and the work of Bliss in the United States. Gaddum wrote a comprehensive report with the emphasis on practical aspects of the experiments and on statistical interpretation, giving several worked examples from the medical and pharmaceutical literature. Bliss published two brief notes in *Science*, and introduced the term *probit*; he followed this up with a series of scholarly articles setting out the maximum likelihood estimation of the model. Both Gaddum and Bliss set standards of estimation; until the 1930s this was largely a matter of ad hoc numerical and graphical adjustment of curves to categorical data.

Both authors adhere firmly to the stimulus and response model of bio-assay, with determinate stimuli and random responses that reflect the distribution of individual tolerance levels. Bliss originally defined the probit (short for 'probability unit') as a convenient *scale* for the equivalent normal deviate, but abandoned this very soon in favour of a different definition which was generally accepted. The equivalent normal deviate of a (relative) frequency f is the solution of $\tilde{Z}$ from

$$f = \frac{1}{\sqrt{2\pi}} \int_{-\infty}^{\tilde{Z}} \exp(-\frac{1}{2}u^2)\mathrm{d}u,$$

and the probit is the equivalent normal deviate increased by 5. This ensures that the probit is almost always positive, which facilitates calculation; at the time this was a common procedure.

The acceptance of the probit method was aided by the articles of Bliss, who published regularly in this field until the 1950s, and by Finney and others (Gaddum returned to pharmacology). The full flowering of this school in bio-assay probably coincides with the first edition of Finney's monograph in 1947. Applications in other fields like economics and market research appear already in the 1950s: Farrell (1954) employed a probit model for the ownership of cars of different vintage as a function of household income, and Adam (1958) fitted lognormal demand curves to survey data of the willingness to buy cigarette lighters and the like at various prices. The classic monograph on the lognormal distribution of Aitchison and Brown (1957) brought probit analysis to the notice of a wider audience of economists.

As far as I can see the introduction of the logistic as a valid alternative is the work of a single person, namely Joseph Berkson (1899–1982), Reed's co-author of the paper on autocatalytic functions mentioned in the last section. Berkson read physics at Columbia, then went to Johns Hopkins for an MD and a doctorate in statistics in 1928. He stayed on as an assistant for three years and then moved to the Mayo Clinic where he remained for the rest of his working life as chief statistician.

In the 1930s he published numerous papers in medical and public health journals. But in 1944 he turned his attention to the statistical methods of bio-assay and proposed the use of the logistic, coining the term 'logit' by analogy with the 'probit' of Bliss. Berkson's advocacy of the logit as a substitute for the probit was tangled by his simultaneous campaign for minimum chi-squared estimation as vastly superior to maximum likelihood. Between 1944 and 1980 he wrote a large number of papers on both issues, often in a somewhat provocative style, and much controversy ensued.

The close resemblance of the logistic to the normal distribution must have been common knowledge among those who were familiar with the logistic; it had been demonstrated by Wilson (1925) and written up by Winsor (1932), another collaborator of Pearl. Wilson was probably the first to use the logistic in bio-assay in Wilson and Worcester (1943), just before Berkson (1944) did so. But it was Berkson who persisted and fought a long and spirited campaign which lasted for several decades.

The logit was not well received by the practitioners of probit analysis, who often regarded it as a cheap and somewhat disreputable device. Thus Aitchison and Brown (1957, p. 72) dismiss the logit in a single sentence by the argument that *'the logistic lacks a well-recognized and manageable frequency distribution of tolerances which the probit curve does possess in a natural way'*. Finney's authoritative textbook of probit analysis kept silent about the logit in the first and second edition of 1947 and 1952; the author only made amends in the third edition of 1971. For a long time no one (not even Berkson) recognized the formidable power of the logistic's analytical properties; one of the first to do so was Cox in his textbook of 1969 which later became Cox and Snell (1989). On the other hand the practice of logit analysis spread more rapidly on the workfloor, in part because it was so much easier to compute.

It should be appreciated that until the advent of the computer and the pocket calculator all numerical work had to be done by hand, that is with pencil and paper, sometimes aided by graphical inspection of 'freehand curves', 'fitted by eye'.† For probit and logit analyses of categorical data there were graph papers with special grids on which a probit or logit curve would appear as a straight line. Wilson (1925) introduces the logistic (or 'autocatalytic') grid, and examples of lognormal paper can be found in Aitchison and Brown (1957) and Adam (1958). Numerical work was supported rather feebly by simple mechanical calculating

---

† We have paid tribute to this tradition in Section 3.5.

machines, driven by hand or powered by a small electric motor, doing addition and multiplication; values of the normal distribution (and of exponentials and logarithms) had to be obtained from printed tables. The *Biometrika* tables of Pearson (1914) were famous. In 1938 a modern set was published by Fisher and Yates. These *Statistical Tables for Biological, Agricultural and Medical Research* from the first carried specially designed tables for probit analysis (with auxiliary tables contributed by Bliss and by Finney). From the fifth edition of 1957 onwards they also included special tables for logits.

In the 1960s the logit gradually achieved an equal footing with the probit, both in academic writings and in practical applications. Probit and logit were also more widely adopted beyond bio-assay, in economics, epidemiology and the social sciences. The close link to tolerance levels or threshold values was dissolved in the more general and abstract representation of the latent regression equation. I believe this was first explicitly formulated by McKelvey and Zavoina (1975) for an ordered probit model of the voting behaviour of US Congressmen. The wider acceptance was greatly helped by the advent of the computer and by the introduction of package routines for the maximum likelihood estimation of both logit and probit models from individual data, as in the BMDP or BIOMEDICAL DATA PROCESSING package of 1977, probably the first to offer this facility. By 1989, when the first edition of the comprehensive textbook of Hosmer and Lemeshow appeared, the use of such routines was taken for granted.

During the 1960s statisticians gradually discovered the superior analytical properties of the logit transformation and its usefulness outside bio-assay. This development was matched by generalizations and extensions of the binary model. The multinomial generalization was first mooted by Cox (1966) and then, independently, by Theil (1969), who immediately saw its potential as a general approach to the modelling of shares. The simple algebra of this generalization opened up a very wide field of applications in economics and other social sciences, and the lead in exploiting its possibilities soon passed to a group of economists led by McFadden. Sophisticated multinomial models were used in empirical work, often undertaken in the course of consultancy, and many theoretical problems were solved in the course of this applied work. The survey article of Amemiya (1981) and the textbooks of Maddala (1983) and Amemiya (1985) made probit and logit models familiar to students of econometrics, and they quickly became familiar research tools of empirical economic research, with the logit much preferred for multinomial analyses. As we have seen in Chapter 8, during the 1990s the wheel turned and the limitation of the IIA property of logit models made advanced market research turn back to the multinomial probit.

9.3 Other derivations

Apart from the route of the stimulus and response model and the adoption of the logistic curve there are several other paths which lead to the logit model. Fisk (1961) derived it as a limiting form of a Champernowne income distribution, and Theil (1969) linked the multinomial version directly to considerations from information theory; but these views were not followed up by their authors or by others. Agresti (1996) hints at a relation with the methods of Rasch, which has some following among psychologists and sociologists; there may well be others of which I am unaware. I shall here conclude with a brief historical note on the method of case–control and on discrete choice, which have both been treated in earlier chapters. Both paradigms have been developed independently and in isolation due to self-imposed separation between scientific disciplines. Logistic regression has largely been the work of biologists, with some later contributions from econometricians; case–control has been developed in medicine, discrete choice in mathematical psychology and in transportation studies. A glance at the bibliographical references of published papers shows that communication among scholars is often confined to a narrow group of kindred minds.

Case–control studies combine two radical shifts in the attitude of medicine to disease. A doctor's natural approach is to treat the patient and study the course of the illness, and epidemiology was revolutionary in widening the outlook to a group of patients and to the contrast of their conditions with those of healthy people. The first examples date from the middle of the 19th century. Secondly, the usual view of a scientific experiment is that the outcome is unknown when it is undertaken. This tradition of prospective or forward-looking observation must be reversed in retrospective samples, collected after the event and even in direct relation with the outcome. As we have argued in Section 6.3, this is the only way of obtaining a sufficient number of cases of specific and rare diseases. Armenian and Lilienfeld (1994) cite a number of examples from the 1920s onward, but the major breakthrough occurred in the 1950s in retrospective studies of the relation between lung cancer and cigarette smoking. Both the method and the substantive findings caused heated controversy among doctors and statisticians. A major contribution establishing the validity of case–control studies was made by Cornfield, himself not a doctor but a self-taught statistician (he read history and started his career at the US Bureau of Labor Statistics); see Cornfield (1951, 1956). The statistical analysis of these samples proceeded from the consideration of risks and relative risks to odds, odd ratios and the log of odds ratios; a vigorous but separate investigation of the statistical issues followed before it was realized that the method is closely related to discriminant analysis and logistic regression. The survey of Breslow (1996) is instructive.

The second independent and distinct route to the logit model and specifically to its multinomial form is from the discrete choice theory of mathematical psychology. We have briefly sketched this relation to the work of Luce and Suppes in the 1950s in Section 8.1. The recognition that it leads to operational statistical models like the general logit specification is very largely due to McFadden (1974), who first read physics at Minnesota (where he obtained a thorough knowledge of differential equations) and who worked as a research assistant in social psychology before he graduated in economics and began his academic career in that field. He soon moved to Berkeley, and there developed the multinomial (conditional) logit model from random choice theory. From 1970 onward he was involved in transportation research consultancy, and soon he and his collaborators began publishing a stream of path-breaking papers on the subject. His views of the role of discrete choice theory are best exemplified by his paper on the nested logit model (McFadden 1981). His personal intellectual development as well as his debt to forerunners and to many of his equally gifted collaborators are set out in his Nobel prize acceptance speech (McFadden 2001).

Bibliography

Abbott, W. S. (1925). A method of computing the effectiveness of an insecticide. *Journal of Economic Entomology* **18**, 205–207.

Adam, D. (1958). *Les Réactions du Consommateur Devant les Prix.* Number 15 in Observation Economique, (Sedes, Paris).

Afifi, A. and Clark, V. (1990). *Computer-aided Multivariate Analysis.* (Lifetime Learning Publications, Belmont, Calif.).

Agresti, A. (1996). *An Introduction to Categorical Data Analysis.* (Wiley, New York).

Aitchison, J. and Brown, J. A. C. (1957). *The Lognormal Distribution.* Number 5 in University of Cambridge, Department of Applied Economics Monographs, (Cambridge University Press, Cambridge).

Aitchison, J. and Silvey, S. (1957). The generalization of probit analysis to the case of multiple responses. *Biometrika* **44**, 131–140.

Amemiya, T. (1981). Qualitative response models: a survey. *Journal of Economic Literature* **19**, 1483–1536.

Amemiya, T. (1985). *Advanced Econometrics.* (Harvard University Press, Cambridge, Mass).

Amemiya, T. and Nold, F. (1975). A modified logit model. *Review of Economics and Statistics* **57**, 255–257.

Armenian, H. K. and Lilienfeld, D. E. (1994). Overview and historical perspective. *Epidemiological Reviews* **16**, 1–5. Special issue on case-control studies.

Bakker, F. M., Klein, M. B., Maas, N. C., and Braun, A. R. (1993). Saturation deficit tolerance spectra of phytophagon mites and their phytoseiid predators on cassava. *Experimental and Applied Acarology* **17**, 97–113.

Baltas, G. and Doyle, P. (2001). Random utility models in marketing: a survey. *Journal of Business Research* **51**, 115–125.

Barnes, P. (2000). The identification of U.K. takeover targets using published historical cost accounting data. *International Review of Financial Analysis* **92**, 147–162.

Beggs, S., Cardell, S., and Hausman, J. (1981). Assessing the potential demand for electric cars. *Journal of Econometrics* **16**, 1–19.

Ben-Akiva, M. and Lerman, S. R. (1987). *Discrete Choice Analysis: Theory and Application to Travel Demand.* (MIT Press, Cambridge, Mass.).

Bergmann, B., Eliasson, G., and Orcutt, G. H. (1980). *Micro Simulation – Models, Methods and Applications*. (Almquist and Wiksell, Stockholm).

Berkson, J. (1944). Application of the logistic function to bio-assay. *Journal of the American Statistical Association* **9**, 357–365.

Berkson, J. (1951). Why I prefer logits to probits. *Biometrics* **7**, 327–339.

Berkson, J. (1980). Minimum chi-square, not maximum likelihood! *Annals of Mathematical Statistics* **8**, 457–487.

Birnbaum, A. (1968). Some latent trait models and their use in inferring an examinee's ability. In *Statistical Theories of Mental Test Scores*, F. M. Lord and M. R. Novick, eds. (Addison–Wesley, Reading, Mass.), 397–479.

Bishop, Y. M. M., Fienberg, S. E., and Holland, P. W. (1975). *Discrete Multivariate Analysis*. (MIT Press, Cambridge, Mass).

Bliss, C. I. (1934a). The method of probits. *Science* **79**, 38–39.

Bliss, C. I. (1934b). The method of probits. *Science* **79**, 409–410.

Borger, J., Kemperman, H., Smit, H. S., Hart, A., van Dongen, J., Lebesque, J., and Bartelink, H. (1994). Dose and volume effects on fibrosis after breast-conservation therapy. *International Journal of Radiation, Oncology, Biology and Physics* **30**, 1073–1081.

Börsch-Supan, A. and Hajivassiliou, V. A. (1993). Smooth unbiased multivariate probability simulators for maximum likelihood estimation of limited dependent variable models. *Journal of Econometrics* **58**, 347–368.

Breslow, N. (1996). Statistics in epidemiology: the case-control study. *Journal of the American Statistical Association* **91**, 14–28.

Breslow, N. and Day, N. (1980). *Statistical Methods in Cancer Research*. (IAC, Lyon).

Centraal Bureau voor de Statistiek. (1997). *Statistisch Jaarboek 1996*. (Centraal Bureau voor de Statistiek, The Hague).

Cornfield, J. (1951). A method of estimating comparative rates from clinical data. *Journal of the National Cancer Institute* **11**, 1269–1275.

Cornfield, J. (1956). A statistical problem arising from retrospective studies. In *Proceedings of the Third Berkeley Symposium on Mathematical Statistics and Probability*, J. Neyman, ed. (University of California Press, Berkeley, Calif.), 135–148.

Cox, D. (1966). Some procedures connected with the logistic qualitative response curve. In *Research Papers in Statistics: Festschrift for J. Neyman*, F. David, ed. (Wiley, London), 55–71.

Cox, D. and Snell, E. J. (1989). *Analysis of Binary Data*, Second edn. (Chapman and Hall, London). First edition, by Cox alone, in 1969.

Cramer, J. S. (1997). Two properties of predicted probabilities in discrete regression models. Tech. rep., Tinbergen Institute. Discussion Paper 97-044/4, download from www.Tinbergen.nl.

Cramer, J. S. (1999). Predictive performance of the binary logit model in unbalanced samples. *The Statistician (Journal of the Royal Statistical Society, series C)* **48**, 85–94.

Cramer, J. S., Franses, P. H., and Slagter, E. (1999). Censored regression in large samples with many zero observations. Tech. rep., Econometric Institute, Erasmus University Rotterdam.

Cramer, J. S. and G.Ridder. (1991). Pooling states in the multinomial logit model. *Journal of Econometrics* **47**, 267–272.

Cramer, J. S. and G.Ridder. (1992). Pooling states in the multinomial logit model: Acknowledgement. *Journal of Econometrics* **51**, 285.

Daganzo, C. (1979). *Multinomial Probit.* (Academic Press, New York).

Davidson, R. and McKinnon, J. G. (1984). Comvenient specification tests for logit and probit models. *Journal of Econometrics* **25**, 241–262.

Davidson, R. and McKinnon, J. G. (1993). *Estimation and Inference in Econometrics.* (Oxford University Press, Oxford).

Davies, A., Wittebrood, K., and Jackson, J. (1997). Predicting the criminal antecedents of a stranger rapist. *Science and Justice* **17**, 161–170.

Debreu, G. (1960). Review of R. D. Luce's 'Individual Choice Behavior'. *American Economic Review* **50**, 186–188.

Domencich, T. A. and McFadden, D. (1975). *Urban Travel Demand: A Behavioral Analysis.* (North Holland, Amsterdam).

Dronkers, J. (1993). Educational reform in the Netherlands: Did it change the impact of parental occupation and education? *Sociology of Education* **66**, 262–277.

Du Pasquier, L.-G. (1918). Esquisse d'une nouvelle théorie de la population. *Vierteljahrsschrift der Naturforschenden Gesellschaft in Zürich* **63**, 236–249.

Efron, B. (1978). Regression and ANOVA with zero-one data: Measures of residual variation. *Journal of the American Statistical Association* **73**, 113–121.

Farrell, M. J. (1954). The demand for motorcars in the United States. *Journal of the Royal Statistical Society, series A* **117**, 171–200.

Fechner, G. T. (1860). *Elemente der Psychophysik.* (Breitkopf und Härtel, Leipzig).

Finney, D. (1971). *Probit Analysis*, Third edn. (Cambridge University Press, Cambridge). First edition in 1947.

Fisher, R. A. and Yates, F. (1938). *Statistical Tables for Biological, Agricultural and Medical Research.* (Oliver and Boyd, Edinburgh).

Fisk, P. (1961). The graduation of income distribution. *Econometrica* **29**, 171–185.

Ford, I., Norrie, J., and Ahmadi, S. (1995). Model inconsistency, illustrated by the Cox proportional hazards model. *Statistics in Medicine* **14**, 735–746.

Franses, P. H. and Paap, R. (2001). *Quantitative Models in Marketing Research.* (Cambridge University Press, Cambridge).

Frerichs, R., Aneshensel, C., and Clark, V. (1981). Prevalence of depression in Los Angeles city. *American Journal of Epidemiology* **113**, 691–699.

Gabler, S., Laisney, F., and Lechner, M. (1993). Seminonparametric estimation of binary-choice models with an application to labor force participation. *Journal of Business and Economic Statistics* **11**, 61–85.

Gaddum, J. H. (1933). *Reports on Biological Standard III. Methods of Biological Assay Depending on a Quantal Response.* (Medical Research Council, London). Special Report Series of the Medical Research Council, no. 183.

Gail, M., Williams, R., Byar, D. P., and Brown, C. (1976). How many controls? *Journal of Chronic Diseases* **29**, 723–731.

Gail, M. H., Wieand, S., and Piantadosi, S. (1984). Biased estimates of treatment effect in randomized experiments with nonlinear regressions and omitted covariates. *Biometrika* **71**, 431–444.

Gaudry, M., Jara-Diaz, S., and Ortuzar, J. (1989). Value of time sensitivity to model specification. *Transportation Research, B* **23**, 151–158.

Gilliatt, R. (1947). Vaso-constriction of the finger following deep inspiration. *Journal of Physiology* **107**, 76–88.

Goldberger, A. S. (1964). *Econometric Theory.* (Wiley, New York).

Gourieroux, C. (2000). *Econometrics of Qualitative Dependent Variables.* (Cambridge University Press, Cambridge). First edition (in French) in 1991.

Greene, W. H. (1992). A statistical model for credit scoring. Tech. rep., Stern School of Business, New York University.

Hajivassiliou, V., McFadden, D., and Ruud, P. (1996). Simulation of multivariate normal rectangle probabilities and their derivatives. *Journal of Econometrics* **72**, 85–134.

Hauck, W. W., Jr and Donner, A. (1977). Wald's test as applied to hypotheses in logit analysis. *Journal of the American Statistical Association* **72**, 851–852.

Hausman, J. A., Abrevaya, J., and Morton, F. S. (1998). Misclassification of the dependent variable in a discrete-response setting. *Journal of Econometrics* **87**, 239–269.

Hausman, J. A. and McFadden, D. (1984). Specification tests for the multinomial logit model. *Econometrica* **52**, 1219–1240.

Hausman, J. A. and Wise, D. A. (1978). A conditional probit model for qualitative discrete decisions recognizing interdependence and heterogeneous preferences. *Econometrica* **46**, 403–426.

Heckman, J. J. (1979). Sample selection bias as a specification error. *Econometrica* **47**, 153–161.

Hewlett, F. (1969). The toxicity to Tribolium Castaneum of mixtures of pyrethrine and piperonylbutoxide: fitting a mathematical model. *Journal of Storing Products Research* **5**, 1–9.

Hill, M. A. (1983). Female labour force participation in developing and developed countries. *Review of Economics and Statistics* **65**, 459–468.

Hosmer, D. W. and Lemeshow, S. (1980). Goodness of fit tests for the multiple logistic regression model. *Communications in Statistics* **A10**, 1043–1069.

Hosmer, D. W. and Lemeshow, S. (2000). *Applied Logistic Regression*, Second edn. (Wiley, New York). First edition in 1989.

Hsiao, C. and Sun, B. H. (1999). Modeling survey response bias – with an analysis of the demand for an advanced electronic device. *Journal of Econometrics* **89**, 15–39.

Imbens, G. W. (1992). An efficient method of moments estimator for discrete choice models with choice-based sampling. *Econometrica* **60**, 1187–1214.

Johnson, N. L. and Kotz, S. (1970). *Distributions in Statistics: Continuous Univariate Distributions.* (Wiley, New York). Two volumes.

Johnson, N. L. and Kotz, S. (1972). *Distributions in Statistics: Continuous Multivariate Distributions.* (Wiley, New York).

Kay, R. and Little, S. (1986). Assessing the fit of the logistic model: a case study of children with the Haemolytic Uraemic syndrome. *Applied Statistics* **35**, 16–30.

Kay, R. and Little, S. (1987). Transformations of the explanatory variables in the logistic regression model for binary data. *Biometrika* **74**, 495–501.

Lachenbruch, P. A. (1975). *Discriminant Analysis.* (Hafner, New York).

Ladd, G. W. (1966). Linear probability functions and discriminant functions. *Econometrica* **34**, 873–885.

Lancaster, K. L. (1971). *Consumer Demand; A New Approach.* (Columbia University Press, New York).

Lancaster, T. (1979). Prediction from binary choice models – A note. *Journal of Econometrics* **9**, 387–390.

Layton, A. P. and Katsuura, M. (2001). Comparison of regime switching, probit and logit models in dating and forecasting US business cycles. *International Journal of Forecasting* **17**, 403–417.

Lee, L.-F. (1982). Specification error in multinomial logit models. *Journal of Econometrics* **20**, 197–209.

Lemeshow, S. and Hosmer, D. W. (1982). A review of goodness of fit statistics for use in the development of logistic regression models. *American Journal of Epidemiology* **115**, 92–106.

Lemeshow, S., Teres, D., Avrunin, J. S., and Pastides, H. (1988). Predicting the outcome of intensive care unit patients. *Journal of the American Statistical Association* **83**, 348–356.

Lennox, C. (1999). Identifying failing companies: a reevaluation of the logit, probit and DA approaches. *Journal of Economics and Business* **51**, 347–364.

Lerman, S. and Manski, C. (1981). On the use of simulated frequencies to approximate choice probabilities. In *Structural Analysis of Discrete Data with Econometric Applications*, C. F. Manski and D. McFadden, eds. (MIT Press, Cambridge, Mass.), 305–319.

Lerman, S. R. and Ben-Akiva, M. (1976). Disaggregate behavioral model of automobile ownership. *Transportation Research Record* **569**, 34–55.

Luce, R. (1959). *Individual Choice Behavior.* (Wiley, New York).

Luce, R. and Suppes, P. (1965). Preferences, utility, and subjective probability. In *Handbook of Mathematical Psychology*, R. Luce, R. Bush, and E. Galanter, eds. (Wiley, New York), 249–410.

Maddala, G. S. (1983). *Limited-dependent and Qualitative Variables in Econometrics.* (Cambridge University Press, Cambridge).

Manski, C. F. (1977). The structure of random utility models. *Theory and Decision* **8**, 229–254.

Manski, C. F. (1985). Semiparametric analysis of discrete response: Asymptotic properties of the maximum score estimator. *Journal of Econometrics* **27**, 313–333.

Manski, C. F. and Lerman, S. R. (1977). The estimation of choice probabilities from choice based samples. *Econometrica* **45**, 1977–1988.

McCullagh, P. and Nelder, J. A. (1989). *Generalized Linear Models.* (Chapman and Hall, London).

McFadden, D. (1974). Conditional logit analysis of qualitative choice behavior. In *Frontiers in Econometrics*, P. Zarembka, ed. (Academic Press, New York), 105–142.

McFadden, D. (1976). Quantal choice analysis: a survey. *Annals of Economic and Social Measurement* **5**, 363–390.

McFadden, D. (1981). Econometric models of probabilistic choice. In *Structural Analysis of Discrete Data with Econometric Applications*, C. F. Manski and D. McFadden, eds. (MIT Press, Cambridge, Mass.), 198–272.

McFadden, D. (1989). A method of simulated moments for estimation of discrete response models without numerical integration. *Econometrica* **57**, 995–1026.

McFadden, D. (2001). Economic choices. *American Economic Review* **91**, 352–370. Nobel prize acceptance speech.

McFadden, D. and Reid, F. (1975). Aggregate travel demand forecasting from disaggregated behavioral models. *Transportation Research Record* **534**, 24–37.

McKelvey, R. D. and Zavoina, W. (1975). A statistical model for the analysis of ordinal level dependent variables. *Journal of Mathematical Sociology* **4**, 103–120.

Meadows, D. H., Meadows, D. L., Randers, J., and Behrens, W. W. (1972). *The Limits to Growth.* (Universe Books, New York).

Menard, S. (1995). *Applied Logistic Regression Analysis.* (Sage Publications, Thousand Oaks, Calif.).

Menard, S. (2000). Coefficients of determination for multiple logistic regression analysis. *American Statistician* **54**, 17–24.

Miner, J. R. (1933). 'Pierre-François Verhulst, the discoverer of the logistic curve. *Human Biology* **5**, 673–689.

Mood, A. M., Graybill, F. A., and Boes, D. C. (1974). *Introduction to the Theory of Statistics*, Third edn. (McGraw-Hill, New York).

Mot, E. and Cramer, J. (1992). Mode of payment in household expenditures. *De Economist* **140**, 488–500.

Mot, E. S., Cramer, J. S., and van der Gulik, E. M. (1989). *De Keuze van een Belaalmiddel.* (Stichting voor Economisch Onderzoek, Amsterdam). In Dutch.

Oosterbeek, H. (2000). Adverse selection and the demand for supplementary dental insurance. *De Economist* **148**, 177–190.

Oosterbeek, H. and Webbink, D. (1995). Enrolment in higher education in the Netherlands. *De Economist* **143**, 367–380.

Orcutt, G. H., Greenberger, M., Korbel, J., and Rivlin, A. M. (1961). *Microanalysis of Socio-economic Systems.* (Harper, New York).

Pakes, A. and Pollard, D. (1989). Simulation and the asymptotics of optimization estimators. *Econometrica* **57**, 1027–1057.

Palepu, K. G. (1986). Predicting takeover targets. *Journal of Accounting and Economics* **8**, 3–35.

Pearl, R. (1922). *The Biology of Death.* (Lippincott, Phildelphia).

Pearl, R. (1927). The indigenous population of Algeria in 1926. *Science* **66**, 593–594.

Pearl, R. and Reed, L. J. (1920). On the rate of growth of the population of the United States since 1870 and its mathematical representation. *Proceedings of the National Academy of Sciences* **6**, 275–288.

Pearl, R. and Reed, L. J. (1923). On the mathematical theory of population growth. *Metron* **5**, 6–19.

Pearl, R., Reed, L. J., and Kish, J. F. (1940). The logistic curve and the census count of 1940. *Science* **92**, 486–488.

Pearl, R., Winsor, C. P., and White, F. B. (1928). The form of the growth curve of the cantaloupe (Cucumis melo) under field conditions. *Proceedings of the National Academy of Sciences* **14**, 895–901.

Pearson, K. (1914). *Tables for Statisticians and Biometricians.* (Cambridge University Press, Cambridge).

Pregibon, D. (1981). Logistic regression diagnostics. *Annals of Statistics* **9**, 705–724.

Pregibon, D. (1982). Score tests in GLIM with applications. In *GLIM82:*

Proceedings of the International Conference on Generalized Linear Models, R. Gilchrist, ed. Lecture Notes in Statistics, vol. 14. (Springer, New York), 87–97.

Prentice, R. and Pyke, R. (1979). Logistic disease incidence models and case-control studies. *Biometrika* **66**, 403–411.

Pudney, S. (1989). *Modelling Individual Choice: the Econometrics of Corners, Kinks and Holes.* (Basil Blackwell, Oxford).

Rao, C. R. (1955). Theory of the method of estimation by minimum chi-square. *Bulletin de l'Institut International de Statistique* **35**, 25–32.

Reed, L. J. and Berkson, J. (1929). The application of the logistic function to experimental data. *Journal of Physical Chemistry* **33**, 760–779.

Ross, S. M. (1977). *Introduction to Probability Models.* (Harcourt Brace Jovanovich, New York).

Ruud, P. A. (1983). Sufficient conditions for the consistency of maximum likelihood estimation despite misspecification of distribution in multinomial discrete choice models. *Econometrica* **51**, 225–228.

Scott, A. J. and Wild, C. J. (1986). Fitting logistic models under case-control or choice based sampling. *Journal of the Royal Statistical Society, series B* **48**, 170–182.

Silber, J. H., Rosenbaum, P. R., and Ross, R. N. (1995). Comparing the contributions of groups of predictors: Which outcome vary with hospital rather than patient characteristics? *Journal of the American Statistical Association* **90**, 7–18.

Stigler, S. M. (1986). *The History of Statistics.* (Harvard University Press, Cambridge, Mass.).

Theil, H. (1969). A multinomial extension of the linear logit model. *International Economic Review* **10**, 251–259.

Thurstone, L. (1927). A law of comparative judgment. *Psychological Review* **34**, 273–286.

Tobin, J. (1958). Estimation of relationships for limited dependent variables. *Econometrica* **26**, 24–36.

Tsiatis, A. A. (1980). A note on a goodness-of-fit test for the logistic regression model. *Biometrika* **67**, 250–251.

Verhulst, P.-F. (1838). Notice sur la loi que la population suit dans son accroissement. *Correspondance Mathématique et Physique, publiée par A. Quetelet* **10**, 113.

Verhulst, P.-F. (1845). Recherches mathématiques sur la loi d'accroissement de la population. *Nouveaux Mémoires de l'Académie Royale des Sciences, des Lettres et des Beaux-Arts de Belgique* **18**, 1–38.

Verhulst, P.-F. (1847). Deuxième mémoire sur la loi d'accroissement de la population. *Nouveaux Mémoires de l'Académie Royale des Sciences, des Lettres et des Beaux-Arts de Belgique* **20**, 1–32.

Wilson, E. B. (1925). The logistic or autocatalytic grid. *Proceedings of the National Academy of Sciences* **11**, 451–456.

Wilson, E. B. and Worcester, J. (1943). The determination of L. D. 50 and its sampling error in bio-assay. *Proceedings of the National Academy of Sciences* **29**, 79. First of a series of three articles.

Windmeijer, F. A. G. (1992). *Goodness of Fit in Linear and Qualitative Choice Models.* (Thesis Publishers, Amsterdam).

Windmeijer, F. A. G. (1995). Goodness-of-fit measures in binary choice models. *Econometric Reviews* **14**, 101–106.

Winsor, C. P. (1932). A comparison of certain symmetrical growth curves. *Journal of the Washington Academy of Sciences* **22**, 73–84.

Wooldridge, J. M. (2002). *Econometric Analysis of Cross Section and Panel Data.* (MIT Press, Cambridge, Mass.).

Xie, Y. and Manski, C. (1989). The logit model and response-based samples. *Sociological Methods and Research* **17**, 283–302.

Yatchev, A. and Griliches, Z. (1985). Specification error in probit models. *Review of Eonomics and Statistics* **67**, 247–258.

Yule, G. U. (1925). The growth of population and the factors which control it. *Journal of the Royal Statistical Society* **138**, 1–59.

Index of authors

Index of subjects

Printed in the United States
By Bookmasters